NATURAL GAS HYDRATES IN FLOW ASSURANCE

NATURAL GAS HYDRATES IN FLOW ASSURANCE

DENDY SLOAN

CAROLYN KOH

AMADEU K. SUM

ADAM L. BALLARD

JEFFERSON CREEK

MICHAEL EATON

JASON LACHANCE

NORM McMULLEN

THIERRY PALERMO

GEORGE SHOUP

LARRY TALLEY

ELSEVIER

AMSTERDAM • BOSTON • HEIDELBERG • LONDON
NEW YORK • OXFORD • PARIS • SAN DIEGO
SAN FRANCISCO • SINGAPORE • SYDNEY • TOKYO

Gulf Professional Publishing is an imprint of Elsevier

G | P
P | ⑂

Gulf Professional Publishing is an imprint of Elsevier
30 Corporate Drive, Suite 400, Burlington, MA 01803, USA
Linacre House, Jordan Hill, Oxford OX2 8DP, UK

Notices

Knowledge and best practice in this field are constantly changing. As new research and experience broaden our understanding, changes in research methods, professional practices, or medical treatment may become necessary.

Practitioners and researchers must always rely on their own experience and knowledge in evaluating and using any information, methods, compounds, or experiments described herein. In using such information or methods they should be mindful of their own safety and the safety of others, including parties for whom they have a professional responsibility.

To the fullest extent of the law, neither the Publisher nor the authors, contributors, or editors, assume any liability for any injury and/or damage to persons or property as a matter of products liability, negligence or otherwise, or from any use or operation of any methods, products, instructions, or ideas contained in the material herein.

Library of Congress Cataloging-in-Publication Data
Natural gas hydrates in flow assurance / E. Dendy Sloan, editor-in-chief.
    p. cm.
    ISBN 978-1-85617-945-4
1. Petroleum pipelines–Fluid dynamics. 2. Natural gas–Hydrates. 3. Offshore oil well drilling–Accidents–Prevention. 4. Gas flow. 5. Natural gas in submerged lands. I. Sloan, E. Dendy, 1944-
    TN879.56.N38 2011
    622'.33819–dc22   2010020759

British Library Cataloguing-in-Publication Data
A catalogue record for this book is available from the British Library.

For information on all Gulf Professional Publishing
publications visit our Web site at www.elsevierdirect.com

ABOUT THE AUTHORS

Adam Ballard is facility engineering team leader for BP's Thunder Horse oil and gas production platform in the deepwater Gulf of Mexico. For 7 years, he has worked for BP in various flow assurance and production roles. Adam received his BS in mathematics from Willamette University and PhD (chemical engineering) from the Colorado School of Mines.

Jefferson Creek holds a BS degree from Middle Tennessee State University, and MS and PhD (1976) degrees from Southern Illinois University at Carbondale. Jeff has worked in phase behavior and fluid analysis since joining Chevron Oil Field Research Company in La Habra, California, in 1977 after a two-year postdoctoral fellowship at UCLA. He continues to work with the flow assurance core team in fluid analysis and phase behavior as integrated with multiphase flow.

Michael Eaton is a senior research engineer with the ExxonMobil Upstream Research Company. His work has focused on the performance of kinetic hydrate inhibitors and the modeling of hydrate-related phenomena. He has a BS and MS in chemical engineering from the Colorado School of Mines, and a PhD in materials science and engineering from the State University of New York at Stony Brook.

Carolyn Koh is associate professor and codirector of the Colorado School of Mines Hydrate Center. She is a fellow of the Royal Society of Chemistry and recipient of the Young Scientist Award of the British Association for Crystal Growth.

Jason Lachance is a flow assurance engineer with the ExxonMobil Upstream Research Company. He has been with the company for 2 years in both research and operation roles. Jason received his BS and MS in chemical engineering from the Colorado School of Mines.

Norm McMullen retired from his senior advisor position at BP America in July 2008. At present, he is working as a flow assurance consultant. Before joining BP in 1997, he was president of PipeSoft Corporation for 4 years and held many positions in Chevron Corporation over the previous 20 years. McMullen holds a BS and MS from the University of Washington, both in mechanical engineering. He is a member of SPE.

Thierry Palermo after a PhD in physical chemistry, entered (IFP) in 1989. He was in charge of the hydrates project and at the head of the complex fluids department for more than 10 years. He then became leader of the flow assurance project, and left IFP in May 2009 to join the production department at TOTAL E&P.

George Shoup is the Paleogene facilities subsea engineering manager for BP. He has worked for BP more than 25 years in various subsea and deepwater production roles. George received his BS degree in mechanical engineering from Rice University and holds MS degrees in both petroleum engineering and mechanical engineering from the University of Tulsa.

Dendy Sloan holds the Weaver Chair in chemical engineering and is the director of the Center for Hydrate Research at the Colorado School of Mines. He is an SPE distinguished lecturer, fellow of the American Institute of Chemical Engineers, and a recipient of the Donald L. Katz Award from the Gas Processors Association.

Amadeu K. Sum is assistant professor in the Chemical Engineering Department at the Colorado School of Mines (CSM) and codirector for the CSM Center for Hydrate Research. He is a current recipient of the DuPont Young Professor Award.

Larry Talley after a PhD in physical chemical physics at the University of California, Riverside, joined Exxon as a research scientist in 1980. He has worked on enhanced oil recovery and flow assurance during his career. He is a coinventor of many kinetic hydrate inhibitors and of a hydrate cold flow process using static mixers.

CONTENTS

5. Artificial and Natural Inhibition of Hydrates 87

Thierry Palermo and Dendy Sloan

LIST OF FIGURES

Figure 1.1
The three repeating hydrate unit crystals and their constitutive cages.

Figure 1.2
Pure-component hydrate formation pressure at 273 K for the first four common hydrocarbons versus logarithm of guest:cage size ratios for optimum cage, from Table 1.1.

Figure 1.3
The relative amounts of methane and water in hydrates compared to the mutual solubility of methane in water.
(From Freer, 2000.)

Figure 1.4
Distance versus time for dimpled, methane hydrate film growth at interface. Looking down through a quartz window through clear methane gas phase, we plot increments of progress in the dimpled methane hydrate phase.
(From Freer, 2000.)

Figure 1.5
Hydrate formation on an emulsified water droplet.

Figure 2.1
Hydrate formation pressures and temperatures (gray region) as a function of methanol concentration in free water for a given gas mixture. Steady-state flowline fluid conditions are shown at distances (indicated as 7 to 50 miles on curve) along the bold black curve.

Figure 2.2
Points of hydrate plug formation in offshore system.

Figure 2.3
Conceptual picture of hydrate formation in oil-dominated system.

Figure 2.4
Integration of CSMHyK into the OLGA flow simulator.

Figure 2.5
Fit of one oil on ExxonMobil flowloop predicts three other oils on University of Tulsa flowloop. There are some similarities of these flowloops from a dimensional and multiphase flow perspective.

Figure 2.6
CSMHyK-OLGA® simulation of hydrate formation in a flowline (*Montesi and Creek, 2006.*)

Figure 2.7
Tommeliten-Gamma lines feeding gas condensate into the Edda platform.

Figure 2.8
Elevation and liquid holdup of Tommeliten-Gamma field service line.

Figure 2.9
Elevation profile of Werner-Bolley field. Numbered squares represent points of P, T, and density (4 only) measurement.

Figure 2.10
Pressure buildup and blockage in the Werner-Bolley line.

Figure 2.11
Hypothesis for hydrate formation from gas-dominated systems. (1) Gas bubbles through low-lying water accumulation. (2) When the subcooling is great enough (e.g., >6 °F), hydrate-encrusted gas bubbles form as the gas exits the water. (3) These hydrate-encrusted bubbles aggregate. (4) Collapse, forming (5) plug downstream of the original water accumulation.

Figure 2.12
Gas-condensate hydrate formation narrowing the flowline channel via wall deposition. The top bent tube blows nitrogen against window to clear it for visibility.
(Courtesy of G. Hatton, Southwest Research Institute.)

Figure 2.13
Simplified condensate flowloop schematic (Nicholas, 2008). Note that the moisture-monitored water concentration difference (inlet minus outlet) is due to solid deposits in the test section, which has pressure and temperature measurements at 40-ft (12.2-m) intervals. Hydrate formation temperatures were achieved at the exit of the cooling section.

Figure 2.14
Difference between water inlet and outlet concentrations, and pressure drop buildup due to hydrate formation in a condensate loop. Note that step changes in pressure drop and concentrations are due to manual flow adjustments to maintain approximately constant flow rates.
(From Nicholas, 2008.)

Figure 2.15
Temperature increases at 40-, 80-, and 120-ft lengths in the condensate flowloop test section due to hydrate wall deposits and propagation downstream. Flow rates show manual step adjustments to maintain approximately constant flow.

Figure 2.16
One-way hydrate plugs form in a condensate flowline. In the figure, the darker shade represents condensate, while white represents hydrate.

Figure 2.17
Hypothesis for hydrate formation in high-water-cut systems, when the water is not totally emulsified, but has an additional free water phase.
(From Joshi, 2008.)

Figure 2.18
Cold flow in which particles are sheared and converted to dry hydrates that will flow with a condensate.
(From Talley et al., 2007.)

Figure 2.19
SINTEF Petroleum Research cold flow concept.
(From Lund et al., 2004.)

Figure 3.1
Hydrate plug removal incident from single-sided depressurization. Inset shows process flow diagram of line and valve. The hydrate plug was initially located at the lower left portion of the line.
(Courtesy of Chevron Canada Resources, 1992.)

Figure 3.2
Hydrated particles accumulated from a flowline in a slugcatcher.
(Courtesy of A. Freitas, Petrobras.)

Figure 3.3
Radial dissociation pictures of three hydrate plug experiments, in which the pipe was opened after 1, 2, and 3 hr of depressurization.

Figure 3.4
Force across hydrate plug due to pressure difference.

Figure 3.5
(a) Hydrate plug projectile eruption from pipeline at bend. (b) High-momentum hydrate plug increases pressure, causing pipeline rupture.
(From Chevron Canada Resources, 1992.)

Figure 3.6
Plug velocity oscillation as a function of time, with dampening by friction and flowline liquids.

Figure 3.7
Effect of volume ratio (downstream to upstream of the plug, shown as parameters with each line) on downstream pressure maximum with time.

Figure 3.8
Optimal pressure ratio versus volume (downstream/upstream) ratio.

Figure 3.9
Safety hazard caused by multiple hydrate plugs that trap intermediate pressure.
(From Chevron Canada Resources, 1992.)

Figure 3.10
Safety hazards of high pressures trapped by hydrates upon heating the center of the hydrate plug.
(From Chevron Canada Resources, 1992.)

Figure 4.1
Typical hydrate equilibrium curve showing increasing temperature.

Figure 4.2
Schematic of Atlantis valve arrangement.

Figure 4.3
Atlantis lateral to Cleopatra gas-gathering system.

Figure 4.4
Solids distribution in jumper.

Figure 4.A1
Basic hydrate plug model.

Figure 4.A2
Hydrate velocity plot. Note that top black curved line locus is outbound from initial position, and bottom black curved line is the return path.

Figure 4.A3
Hydrate plug location versus time. Note that very little time is required to move this rather large mass. In reality, rupture would occur, the plug fragmented, and pipework and structural supports damaged.

Figure 4.A4
Chamber pressure versus time. Very quickly the pipework receives a high-pressure pulse that is likely to rupture the pipe wall near the end of the low-pressure chamber. In this case it would be a closed valve or another hydrate plug.

Figure 5.1
Hydrate formation pressures and temperatures (gray region) as a function of methanol concentration in free water for a given gas mixture. Flowline fluid conditions are shown at distances along the bold black curve.

Figure 5.2
Molecular models of (a) methanol and (b) ethylene glycol. The black spheres represent carbon atoms, whites hydrogen, and gray greys are oxygen.

Figure 5.3
Subcooling temperature chart. Note that kinetic hydrate inhibitors (KHIs) are ranked by the degree of subcooling (ΔT) that they can provide below the equilibrium temperature T_{eq} for a given pressure.

Figure 5.4
Repeating chemical formulas for four kinetic hydrate inhibitors. Every line angle in the figure represents a CH_2 group. The upper horizontal angular line with a repeated parenthesis "$( )_{x \text{ or } y}$" in each structure suggests that the monomer structure is repeated "x or y" times to obtain a polymer.

Figure 5.5
Conceptual diagram of hydrate kinetic inhibition mechanism.

Figure 5.6
Conceptual picture of hydrate formation in an oil-dominated system.

Figure 5.7
Photograph of a hydrate particle grown in the presence of sorbitan monolaurate (Span-20) at left, and measured forces at right between two such particles as a function of distance. *(From Taylor, 2006.)*

Figure 6.14
System pressure (MPa) versus temperature (°C) in an isochoric T^{eq} run.

Figure 6.15
Observed hydrate formation via a temperature spike and volume drop in an isobaric autoclave.

Figure 6.16
Observed hydrate formation via a temperature spike and pressure drop in an isochoric autoclave.

Figure 6.17
KHI hold-time results in a CO_2- and H_2S-containing autoclave compared to CO_2-containing miniloop results.

Figure 6.18
South Pass 89A in the Gulf of Mexico.

Figure 6.19
Process and instrumentation diagram for high-pressure miniloop 2.

Figure 7.1
Subsea system and potential location of hydrate blockage.

Figure 7.2
A typical production system with some flow assurance problems labeled to indicate the complexity of the entire flow assurance endeavor.

Figure 7.3
Schematic of dry tree system in which the fluids are still warm at the platform choke, so that separation, drying, and dehydration can be done with a minimum of hydrate inhibition. Such a system is expensive, requiring short connections to the reservoir and fewer wells per platform.

Figure 7.4
Schematic of wet tree design system in which more and longer flowlines are attached to a single platform. Wet systems generally require more hydrate inhibition precautions than the dry tree system shown in Figure 7.3.

Figure 7.5
Generic well system with methanol (MeOH) as a hydrate inhibitor.

Figure A.1
Elevation profile of the Werner-Bolley flow line with locations of temperature-pressure ports indicated. Shaded portion of plot shows most frequent site of plug formation.

Figure A.2
The temperature-pressure profile of the flow line plotted on the hydrate stability curve (without methanol). Numbers denote positions of Bell Holes for P,T measurement as shown in Figure A.1.

Figure A.3
Werner-Bolley data from Test 1, including pressure differentials (with respect to point 5) and gamma densitometer at point 4. P indicates pigging of the line to remove liquids (pig trip calculated at 45 minutes), including methanol (Deepstar CTR 5902-1, pg 3).

Figure A.4
Werner-Bolley data from Test 4, including pressure differentials (with respect to point 5) and gamma densitometer at point 4. P indicates pigging of the line to remove liquids (pig trip calculated at 45 minutes), including methanol (Deepstar CTR 5902-1, pg 3).

Figure A.5
(a) Tommeliten Subsea Tieback to the Edda Platform from Austvik et al. 1997 and (b) the Approximate Topology of the Tommeliten Service Line.

Figure A.6
Oil and water accumulations where the flowline elevation changes. Note that hydrates formed at about 7 and 8 km from the wellhead, with a conjunction of water accumulation and subcooling.

Figure A.7
Matterhorn 10" gas export pipeline.

Figure A.8
Hydrate slurries, top two photographs. Bottom two photographs: (left) hydrate in pig receiver and (right) in barrel.

Figure A.9
The line pressure and temperature operating conditions imposed on a plot of the hydrate formation line (labelled Hydrate, 0% MeOH) and the vapor-liquid phase envelope (0, 1, 2, 4, 6% liquids). This plot shows the line to be sub-cooled almost 30°F at the SCR TDP, with approximately 1% hydrocarbon liquid condensed.

Figure A.10
In the Merganser Field above, the system was restarted about an hour after shut-in. The MEG inhibitor was inadvertently not begun with well restart. After restart, the pressure gradually increased on ramp-up, indicating the formation of a hydrate deposit. Upon the pressure buildup the choke rate was reduced and MEG injection was begun, resulting in a hydrate dissociation. After the pressure had reduced the system production was increased, this time with MEG injection and normal pressures, without hydrate buildup.

Figure A.11
Jubilee 4 downstream choke pressure (psia) versus time. Note that the pressure first increased at 2:35 pm and the line plugged at 3:05 pm.

Figure A.12
Due to the pressure drop, it was determined that the most likely plug location was either the 6 inch jumper or sled or manifold, 1.4 miles downstream of the tree. The temperature dropped from 120°F to 44°F over this 1.4 mile length.

Figure A.13
Downstream pressure in Jubilee 4 versus time. Spike in pressure shows methanol injection points.

Figure A.14
West Boomvang subsea layout.

Figure A.15
Pressure and temperature transients on hydrate plug formation in D2 flowline.

Figure A.16
Orientation of Leon Platform relative to export facility.

Figure A.17
Liquid and water holdup at shut-in for (left) full and (right) reduced production.

Figure A.18
Flowline topology (bottom), temperature (middle), and hydrate equilibrium temperature (top) as a function of line length.
(Courtesy of Leon Field, © Chevron, 2009.)

Figure A.19
CSMHyK-OLGA$^{(\circledR)}$ prediction of increasing hydrate fraction at the point of maximum water holdup.

Figure A.20
CSMHyK-OLGA$^{(\circledR)}$ simulation indicating hydrate plug (rapid increase in pressure) at point of maximum viscosity.

PREFACE

This is a condensed, updated version of earlier works to enable the flow assurance engineer to quickly answer seven questions:

(1) How do hydrate plugs form?

(2) How can hydrate plugs be prevented from forming?

(3) How to deal safely with hydrate plugs?

(4) How to remove a hydrate plug once it has formed?

(5) How can kinetic inhibitors be certified?

(6) What is the mechanism for naturally inhibited oils?

(7) What are industrial hydrate case studies?

Our focus is offshore systems, from the reservoir to the platform, because these lines are the most inaccessible and, thus, the most problematic. However, almost all of the content can be applied to onshore processes and export lines from platforms.

The intent was to combine eight industrial flow assurance perspectives (from British Petroleum [BP], Chevron, ExxonMobil, and the Total Petroleum) with three perspectivs from the Colorado School of Mines to enable resolution of hydrate design and operating problems. In a few pages the coauthors encapsulated knowledge from their careers to provide a basis for advancement by flow assurance engineers.

The trend over the last decade has focused on risk management to manage hydrates in field developments. Thus, the technical perspective of hydrate flow assurance is changing significantly, from avoidance to risk management. While industry previously chose to avoid having transportation equipment operate in the hydrate formation region of pressure and temperature (i.e., by inhibitor injection), a change in that earlier concept is to allow hydrate particles to form, while preventing hydrate plug formation. As this book illustrates, both economic and technical incentives are provided by adding new hydrate risk management tools to the existing tools of hydrate avoidance.

Our intention was to combine the practical experience of industry together with the concepts generated in academia, to state in this concise volume the basics of the new risk-management methods. We gratefully acknowledge the flow assurance engineers who contributed to enable this book: John Abrahamson, Alex Alverado, Guro Aspenes, Torstein Austvik, Ray Ayres, Jim Bennett, Gary Bergman, Phaneendra Bollavaram, John

Boxall, Jep Bracey, Ricardo Carmargo, Richard Christiansen, Jim Chitwood, Mike Conner, Chris Cooley, Simon Davies, Emmanuel Dellecasse, Laura Dieker, Mark Ehrhardt, Havard Eidsmoen, Douglas Estanga, Erik Freer, John Fulton, Jim Grant, David Greaves, Kathy Greenhill, Arvind Gupta, Ronny Hanssen, Greg Hatton, Chris Haver, Blake Hebert, Pål Hemmingsen Prof. Jean-Michel Herri, Scott Hickman, Nikhil Joshi, Sanjeev Joshi, Bob Kaminsky, Moussa Kane, Sam Kashou, Aftab Khokhar, Dean King, Keijo Kinnari, Veet Kruka, Roar Larsen, Joe Lederhos, Emile LePorcher, Jin-Ping Long, Susan Lorimer, Taras Makogon, Patrick Matthews, Ajay Mehta, Dave Miller, Kelly Miller, Pierre Montaud, Alberto Montesi, Julie Morgan, Alex Mussumeci, Bob Newton, Lewis Norman, Phil Notz, Bill Parrish, David Peters, Mike Phillips, David Qualls, Kevin Renfro, Kevin Rider Jeremy Rohan, Laura Rovetto, Allan Rydahl, Mike Scribner, Prof. Sami Selim (deceased), Don Shatto, Rob Sirco, Prof. Johan Sjöblom, Chris Smith, Mark Stair, Tim Strobel, Dan Subik, Siva Subramanian, Craig Taylor, Saadedine Tebbal, Troy Trosclar, Doug Turner, Mike Volk, Sung Oh Wang, and Nick Wolf.

Houston, Texas, Pau, France, and Golden, Colorado
July 2010

CHAPTER ONE

Introduction

Dendy Sloan

Contents

1.1 WHY ARE HYDRATES IMPORTANT?

Oil and gas facilities are expensive, but offshore oil and gas facilities are very expensive. Frequently the cost of an offshore project exceeds US$ 1billion, with wells and flowlines comprising 39% and 38%, respectively, of the total cost (Forsdyke, 2000, p. 4). It is rare for a single company to assume the total risk of such a development, so partnerships are the norm. Field and project partners change from one project to the next, taking information with them to new projects, so that a common body of flow assurance knowledge is becoming available.

In a survey of 110 energy companies, flow assurance was listed as the major technical problem in offshore energy development (Macintosh, 2000, citing Welling and Associates, 1999). Beginning a September 24, 2003 flow assurance forum, James Brill (University of Tulsa) discussed the need for a new academic discipline called "flow assurance." Such a question, presented to an audience of 289 flow assurance engineers, would not have been considered a decade previously, when the flow assurance community totaled a few dozen people. The term "flow assurance" was not coined until 1995, with a modification of the term "Procap" used by Petrobras. Yet the statement of Brill's question indicates the importance of flow assurance.

Safety is perhaps a more important reason for understanding hydrate blockages. Every few years, somewhere in the world a major injury occurs

Natural Gas Hydrates in Flow Assurance
ISBN 978-1-85617-945-4
DOI: 10.1016/B978-1-85617-945-4.00001-7

and/or major equipment damage is done, due to hydrates. After the first two chapters on fundamental structures, and how hydrate plugs form, Chapter 3 deals with hydrate safety, providing guidelines and case studies. The remainder of the book deals with hydrate industrial practices and safe removal of plugs.

The above two areas, safety and flow assurance, are the major topics of this book. The principle of the book is that experience is a better guide to reality than is theory, as reflected by most of the coauthors' careers and the case studies herein.

1.2 WHAT ARE HYDRATES?

Natural gas hydrates are termed "clathrates" or inclusion compounds. This means that there is a network of cages of water molecules that can trap small paraffin guest molecules, such as methane, ethane, and propane.

Of the three common hydrates structures (known as structures I, II, and H), only structures I and II are typically found in oil and gas production and processing. Yet the principles presented in this book for structures I and II apply equally well to structure H. A detailed discussion of structure is given elsewhere (Sloan and Koh, 2008). The following four structure rules of thumb in this introductory chapter are applied in safety and flow assurance:

- Fit of the guest molecule within the host water cage determines the crystal structure.
- Guest molecules are concentrated in the hydrate, by a factor as high as 180.
- Guest:cage size ratio controls formation pressure and temperature.
- Because hydrates are 85 mole % water and 15 mole % gas, gas–water interfacial formation dominates.

At the chapter conclusion, the previous principles are illustrated showing the concept of how hydrates form on a water droplet.

1.2.1 Hydrate Crystal Structures

Figure 1.1 shows the three hydrate unit crystal structures, the smallest crystal unit that repeats itself in space. It is important to review these structures to obtain a basic understanding at the microscopic level, which impacts macroscopic hydrate plugs.

The three rightmost structures in Figure 1.1 are composed of cages, but particularly a basic cage, the 5^{12}, forms as a building block for all three. The 5^{12} cage is composed of 12 pentagonal faces, formed by water

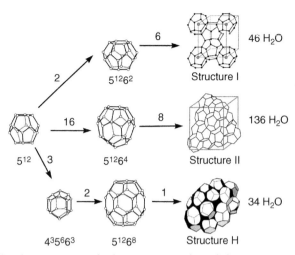

Figure 1.1 The three repeating hydrate unit crystals and their constitutive cages.

molecules that are hydrogen-bonded to each other, with an oxygen at each vertex. Inside the 5^{12} free diameter (5.1 Å) is a hydrocarbon molecule like methane (4.36 Å diameter), which effectively props the cage open. There are no chemical bonds between a cage and a guest molecule; rather the presence of the guest keeps the cage open. Without most of the cages filled, hydrogen-bonded hydrate structures collapse and do not exist in water.

When the 5^{12} cage is connected to others like it via the vertices, a body-centered cubic crystal of 5^{12} cages forms, called hydrate structure I, which exists primarily outside the pipeline, in nature. However, because the 5^{12} cavities alone cannot fill space without strain on the hydrogen bonds, the bond strain is relieved by the inclusion of hexagonal faces to form connecting $5^{12}6^2$ cages, with both the 12 original pentagonal faces and two additional hexagonal, strain-relieving faces.

The free diameter of the $5^{12}6^2$ cage is somewhat larger (5.86 Å) and can contain molecules the size of ethane (5.5 Å diameter), typically the second most common component of natural gas. Methane can fit in the $5^{12}6^2$ cage also, when hydrates are formed from pure methane gas. But methane is too small to prop open the $5^{12}6^2$ effectively, so when mixtures of methane and ethane form structure I (sI), the ethane molecules reside in the $5^{12}6^2$ cages because ethane is too large for the 5^{12} cage. In mixtures of methane and ethane, methane resides mostly in the 5^{12} cages and a small number of the $5^{12}6^2$ cages. In some circumstances, methane and ethane can combine to form structure II (sII) (Sloan and Koh, 2008, chapter 2).

In sum, two 5^{12} cages and six $5^{12}6^2$ cages with 46 water molecules, comprise the sI repeating unit crystal shown in Figure 1.1. Structure I is found mostly in nature because methane is the major component of most hydrates found outside the pipeline. Figure 1.1 shows the sI unit crystal fits a cube 12 Å on a side.

When a larger hydrocarbon, such as propane (6.3 Å diameter), is present in a gas, the propane molecule is too large to be contained in the $5^{12}6^2$ cage, so a larger $5^{12}6^4$ cage (6.66 Å free diameter) forms around larger molecules, such as propane and i–butane (6.5 Å diameter). The $5^{12}6^4$ cage, with twelve pentagonal and four hexagonal faces, is the large cage that relieves hydrogen bond strain when the 5^{12} basic building blocks are connected to each other via their faces. Again the 5^{12} cages cannot fill space, when the 5^{12} are connected to each other, but this time by their pentagonal faces.

The combination of 16 small 5^{12} cages with 8 large $5^{12}6^4$ cages forms the sII unit crystal shown in Figure 1.1, incorporating 136 water molecules in this smallest repeating structure. The sII hydrates are typically found in gas and oil operations and processes, and will be the major concern of this book. The diamond lattice of sII is in a cubic framework that is 17.1 Å on a side.

Still larger molecules such as normal pentane (9.3 Å diameter) cannot fit in any sI or sII cages and are excluded from the hydrate structures of concern to us. The structure H crystals are seldom found in artificial or in natural processes, so we will not deal with them here.

1.3 FOUR RULES OF THUMB ARISING FROM CRYSTAL STRUCTURE

The above discussion leads to the first rule of thumb of hydrate structures:

1. *The fit of the guest molecule within the water cage determines the crystal structure.*

   Consider Table 1.1 that shows the size ratios (or the fit) of the first five common hydrocarbon gases (CH_4, C_2H_6, C_3H_8, i-C_4H_{10}, and n-C_4H_{10}) in the two cages of sI (5^{12} and $5^{12}6^2$) and the two cages of sII (5^{12} and $5^{12}6^4$).

   Note that in Table 1.1, the superscripted symbol "F" indicates the cage occupied by a pure gas, so that pure CH_4 and C_2H_6 are indicated to form sI, while pure C_3H_8 and i-C_4H_{10} form sII. Table 1.1 provides three guidelines for hydrate structure and stability:

Table 1.1 Ratio of guest molecule diameters[a] to cage diameters[b] for five natural gas hydrate formers

| Guest hydrate former | | Molecular diameter/cage diameter | | | |
| | | Structure I | | Structure II | |
Molecule	Diameter[a] (Å)	5^{12}	$5^{12}6^2$	5^{12}	$5^{12}6^4$
CH_4	4.36	0.86[F]	0.74[F]	0.87	0.66
C_2H_6[a]	5.5	1.1	0.94[F]	1.1	0.84
C_3H_8	6.28	1.2	1.1	1.3	0.94[F]
i-C_4H_{10}	6.5	1.4	1.1	1.3	0.98[F]
n-C_4H_{10}	7.1	1.4	1.2	1.4	1.1

An "F" superscript indicates the cage(s) occupied by a pure gas.
[a]Molecular diameters obtained from von Stackelberg and Muller (1954), Davidson (1973), or Davidson et al. (1984, 1986).
[b]Free diameters of cages from x-ray diffraction, minus a value for intrusion of water molecules into each cage.

a. **Molecules that are too large for a cage will not form in that cage as a single guest.** This principle is illustrated by the case of n-C_4H_{10}, which is too large (guest:cage size ratios above 1.0) for every cage and so forms neither sI nor sII as a single guest. However, smaller hydrocarbons will fit in at least one of the four cages, with a fit that determines the structure formed. Ethane, propane, and i-butane fit the larger cages of their structures; pure ethane forms in the $5^{12}6^2$ cage of sI, while propane and i-butane each form in the $5^{12}6^4$ cage of sII, leaving the smaller cages vacant in each case.

b. **The second guideline from Table 1.1 is the optimal size for the guest:cage diameter ratio is between 0.86 and 0.98.** For ratios less than 0.8 the guest does not lend much repulsive stability to the cage. For example, methane props open the $5^{12}6^2$ (size ratio of 0.74) better than it does the $5^{12}6^4$ (size ratio of 0.66), while the ratio is almost the same for methane in the 5^{12} of sI and sII (0.86 and 0.87, respectively), so sI is the stable structure for pure methane hydrate.

c. **Due to the similar ratios for methane in the 5^{12} of sI and sII, the controlling factor is the fit of the large cavity, as shown in item 1b above.** However, if natural gases contain any amount of C_3H_8 and/or i-C_4H_{10}, those larger molecules will only be able to fit into the $5^{12}6^4$ of sII. Since the size ratio of methane in the 5^{12} of sI and sII is almost identical, the presence of any amount of a larger, common molecule (e.g., C_3H_8 or i-C_4H_{10}) will convert the hydrate structure to sII in almost all instances. It is relatively rare to find sI hydrates in oil and gas production due to the common presence of larger molecules.

2. *Hydrates concentrate energy equivalent to a compressed gas.*

If all cages were filled in sI or sII, the guest molecules would be much closer together than in the gas phase at ambient conditions. In fact, hydrate concentrates the gas volume by as much as a factor of 180, relative to the gas volume at 273 K and 1 atmosphere. This concentration has the energy density of a compressed gas, but hydrates have only 42% of the energy density of liquefied methane.

3. *The guest:cage size ratio controls hydrate formation pressure and temperature.*

Figure 1.2 shows the hydrate formation pressure at 273 K as a function of the guest:cage size ratio, for the optimum cage size stabilized by the first four hydrocarbons in Table 1.1 (5^{12} for CH_4, $5^{12}6^2$ for C_2H_6, $5^{12}6^4$ for C_3H_8, and $5^{12}6^4$ for i-C_4H_{10}). As a molecule better fits the cage, the formation pressure decreases. Methane is a relatively poor fit (0.86) in the 5^{12}, so methane's hydrate formation pressure is high (2.56 MPa). In contrast, i-C_4H_{10} is a good fit (0.98) in the $5^{12}6^4$ cage yielding a low formation pressure (0.133 MPa).

The above rules of thumb connect the pressure–temperature stability of hydrates to their crystal structure. Normally, however, natural

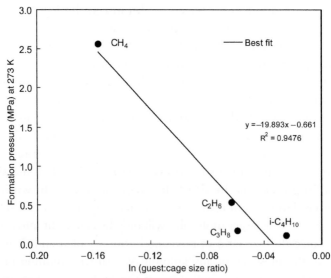

Figure 1.2 Pure-component hydrate formation pressure at 273 K for the first four common hydrocarbons versus logarithm of guest:cage size ratios for optimum cage, from Table 1.1.

gases are composed of a number of components, including acid gases CO_2 and H_2S, relative to the simple one- and two-component gases for which the above rules of thumb apply.

To determine the hydrate stability conditions for more complex gases, the reader is encouraged to use a commercial computer program, such as Multiflash®, PVTSim®, DBRHydrate®, HWHyd, or CSMGem, to determine the hydrate formation conditions of pressure and temperature. Simpler hand calculation methods, together with CSMGem and a user's manual are provided in Sloan and Koh (2008).

4. *Because hydrates are 85 mole % water and 15 mole % gas, hydrate usually forms at the gas–water interface.*

Hydrates contain much higher concentrations of hydrocarbon and water than are normally found in a single phase. At ambient conditions, every 10,000 molecules of liquid water dissolve only eight molecules from a gaseous methane atmosphere. Similarly, 1000 molecules of gaseous methane will only have a single molecule of liquid water vaporized into the gas. If we consider the solubility rule of thumb, "like dissolves like," we must conclude that methane and water are very dissimilar molecules due to their mutual insolubility.

In all three hydrate crystal structures, however, the molecular ratio of hydrocarbon to water is very high (15:85) when all cages are filled. These relative concentrations—in hydrate, in the gas, and in the water phases—are shown in Figure 1.3.

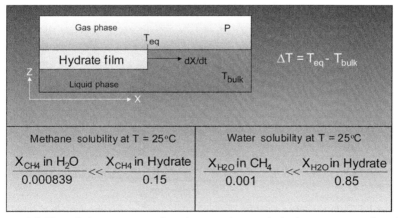

Figure 1.3 The relative amounts of methane and water in hydrates compared to the mutual solubility of methane in water. *(From Freer, 2000.)*

Hydrates will likely form at the methane–water interface. The low solubility of methane in water suggests that only a small amount of hydrate cages will form in the body of the liquid, as there are insufficient methane molecules to occupy many hydrate cages in the liquid water. Similarly the small amount of water vaporized in the methane gas provides only a few hydrate cages forming in the body of the gas; there are too few water molecules to make significant amounts of the host crystal structure in the bulk gas.

Hydrates normally form at the hydrocarbon–water interface, as a consequence of the mismatch between (1) the requirement of high concentrations of both hydrate components and (2) the mutual insolubility (low concentrations) of hydrocarbons and water. Hydrates form at the interface between the two phases, rather than the body of liquid water or the body of gaseous methane.

When one looks down through a quartz glass window into a methane-pressurized cell of water when hydrates form, one sees through the clear methane gas, giving a picture like that shown in the inset of Figure 1.4. Hydrate, the dimpled phase in the photo insert of Figure 1.4, forms from left to right across the clear gray water interface, shown on the right.

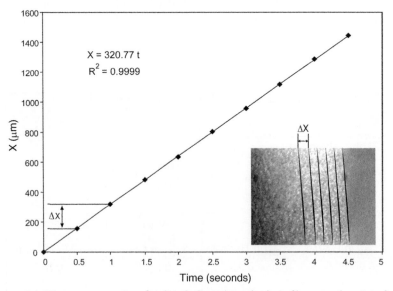

Figure 1.4 Distance versus time for dimpled, methane hydrate film growth at interface. Looking down through a quartz window through clear methane gas phase, we plot increments of progress in the dimpled methane hydrate phase. *(From Freer, 2000.)*

With a fast camera, one can take time-lapse pictures, and plot the linear progress of the film at time and distance increments (Figure 1.4). Regression of the distance with time, shows that the hydrate film grows fairly rapidly, at about 1 mm every 3 sec (Freer, 2000).

The hydrate film grows across the water–gas interface very rapidly. The initial thickness of the hydrate film is small, between 5 and 30 μm (Taylor et al., 2007). The solid phase forms a barrier between the methane gas and the water phases, so that the initial hydrate film prevents further contact of the gas and water, and the hydrate film thickens very slowly. The solid hydrate film formed at the vapor–liquid interface controls the subsequent rates of hydrate formation.

1.4 CHAPTER SUMMARY APPLICATION: METHANE HYDRATE FORMATION ON AN EMULSIFIED WATER DROPLET

How do hydrates form on a water droplet that is emulsified in an oil phase? This question is at the heart of many hydrate applications, such as stabilized cold flow and anti-agglomerant inhibitors, further discussed in Chapters 2 and 5.

When a water droplet is emulsified in an oil phase, hydrates form at the oil–water interface. One conceptual formation picture for hydrate formation is shown in Figure 1.5.

In Figure 1.5 a grey emulsified water droplet, typically around 40 μm in diameter, is shown on the left in an oil environment. Emulsified water droplets occur in most oil systems, caused by a combination of surface chemistry which lowers the interfacial tension, and shear caused by flow, to enable the droplet stability. (Different conceptual pictures to Figure 1.5 for hydrate formation in other systems, such as gas-dominated and high-water-cut systems, are given in Sections 2.2.2 through 2.2.4.)

Just to the left of Point 1 in Figure 1.5, hydrate has begun to form at the water-oil interface, from oil-dissolved, small (gas) molecules which

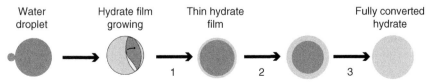

Figure 1.5 Hydrate formation on an emulsified water droplet.

fit into the hydrate cages. The initial hydrate formation thickness is 3000 to 18,000 times the unit crystal length of sII.

This initial formation process illustrates three rules-of-thumb in this chapter:

1. The gas molecules must be small enough to fit into the cages, so only molecules smaller than pentane form hydrates. Oil molecules are much too large to fit into the hydrate cage.

2. The concentration of the guest molecules determines the formation pressure and temperature, as a function of how well the mixed guests fit inside the hydrate cages.

3. Formation occurs at the oil–water interface from small guest molecules dissolved in the oil from the gas phase.

To the right of Point 1 in Figure 1.5, the hydrate film rapidly (1 mm/ 3 sec) covers the entire droplet surface. The film thickness (5–30 μm) slows the hydrate growth by separating the oil and water phases, preventing further contact of the hydrate guest and host components.

To the right of Point 2 in Figure 1.5, the hydrate film slowly thickens, caused by diffusion of the components through the hydrate film. The small guest molecules are so sparingly soluble in the water droplet that they must come from the external oil phase to convert more of the water droplet into hydrate. The slow conversion of the inner droplet is a shrinking core model of hydrate formation.

To the right of Point 3 in Figure 1.5 the entire water droplet has converted to hydrate. The total conversion of the inner droplet is a strong function of how large the droplet is. Large droplets (>40 μm) may require hours to days to fully convert because the growth is hindered by the hydrate film.

In comparison, small (<15 μm) water droplets may convert entirely with the first film formation at the water–oil interface, where the film is 5 to 30 μm thick. In that case, no significant amount of water remains in the droplet, causing the resulting hydrate particle to be "dry" (with little/no adhesion between hydrate particles). These dry hydrate particles are one of the hypothesized mechanisms of stabilized cold flow, discussed in Chapter 2.

Perhaps the most important aspect of hydrate formation is plug prevention and removal. The most important aspect of plug removal is safety. We first consider where, when, and how hydrate plugs form in Chapter 2 to gain insight into which potential operations would facilitate safe hydrate plug removal.

REFERENCES

Davidson, D.W., 1973. Clathrate Hydrates. In: Franks, F. (Ed.), Water: A Comprehensive Treatise Plenum Press, New York, pp. 115-234.

Davidson, D.W., Handa, Y.P., Ratcliffe, C.I., Tse, J.S., Powell, B.M., 1984. The Ability of Small Molecules to Form Clathrate Hydrate Structure II. *Nature* 311, 142.

Davidson, D.W., Handa, Y.P., Ratcliffe, C.I., Ripmeester, J.A., Tse, J.S., Dahn, J.R., Lee, F., Calvert, L.D., 1986. Crystallographic Studies of Clathrate Hydrate, Part I. Mol. Cryst. Liq. Cryst. 141.

Forsdyke, I., 2000. Integrated Flow Assurance Overview and Strategy Report. DeepStar Report CTR 4211-1.

Freer, E., 2000. Methane hydrate formation kinetics. Master's thesis, Colorado School of Mines.

Welling and Associates, 1999. Survey cited by Macinosh, Flow Assurance Still Leading Concern Among Producers. *Offshore*, 60(10), October 2000.

Sloan, E.D., Koh, C.A., 2008. *Clathrate Hydrates of Natural Gases*, third ed. Taylor and Francis, Boca Raton, FL.

Taylor, C.J., Miller, K.T., Koh, C.A., Sloan, E.D., 2007. Macroscopic investigation of hydrate film growth at the hydrocarbon/water interface. *Chemical Engineering Science* 62, 6524.

von Stackelberg, M., Müller, H.R., 1954. Struktur und Raumchemie. *Z. Electrochem.* 58, 25.

CHAPTER TWO

Where and How Are Hydrate Plugs Formed?

Dendy Sloan, Jefferson Creek, and Amadeu K. Sum

Contents

The objective of this chapter is to describe where and how hydrates form, to enable flow assurance or plug prevention, as well as safe dissociation when plugs do form. This chapter provides background for the safety, prevention, and remediation chapters that follow. The chapter builds on the structural information in Chapter 1, which would be helpful to read beforehand.

2.1 WHERE DO HYDRATES FORM IN OFFSHORE SYSTEMS?

Notz (1994) provided a case study of hydrate formation, shown in the pressure–temperature diagram of Figure 2.1 for a deepwater flowline fluid. To the right, in the white region at high temperature, hydrates will not form and the system will exist in the fluid (hydrocarbon and water) region. However, hydrates will form in the gray region at the left of the line marked "Hydrate Formation Curve," so hydrate prevention measures should be taken. Note that the hydrate formation curve varies with every

Natural Gas Hydrates in Flow Assurance
ISBN 978-1-85617-945-4
DOI: 10.1016/B978-1-85617-945-4.00002-9

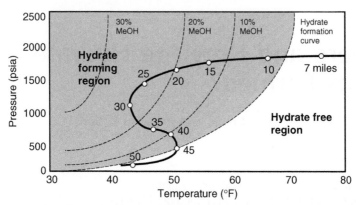

Figure 2.1 Hydrate formation pressures and temperatures (gray region) as a function of methanol concentration in free water for a given gas mixture. Steady-state flowline fluid conditions are shown at distances (indicated as 7 to 50 miles on curve) along the bold black curve.

reservoir fluid and may be predicted with any of the commercial programs listed in Chapter 1.

In Figure 2.1 at 7 mi from the subsea wellhead, the steady-state, flowing stream retains some reservoir heat, thus preventing hydrates. The ocean cools the flowing stream and at about 9 miles a unit mass of flowing gas and associated water enters the hydrate region to the left of the hydrate formation curve, remaining in the uninhibited hydrate area until mile 45. By mile 30, the temperature of the pipeline system is a few degrees higher than the deep ocean temperature (~39 °F), so that approximately 23 wt% methanol is required in the free water phase to shift the hydrate formation region to the left of flowline conditions to prevent hydrate formation and blockage. As vaporized methanol flows along the pipeline from the injection point at the wellhead in Figure 2.1, methanol dissolves into any produced water or water condensed from the gas.

Hydrate formation and accumulation occurs in the free water, usually just downstream of water accumulations, where there is a change in flow geometry (e.g., a bend or pipeline dip along an ocean floor depression) or some nucleation site (e.g., sand, weld slag, etc.). Hydrate formation occurs at the interface of the aqueous liquid that generally contains the highest methanol concentration, rather than in the bulk vapor or oil/condensate.

Hydrate plugs occur during transient and abnormal operations such as on start-up, or restart following an emergency, operational shut-in, or when uninhibited water is present due to dehydrator failure or inhibitor injection failure, or when cooling occurs with flow across a valve or restriction. Hydrate plug formations do not occur during normal flowline operation (and in the absence of unforeseen failures) due to system design for flow assurance. Typically, oil-dominated systems with a higher heat capacity to retain the reservoir temperature than gas systems are less prone to hydrate plug formation. Many oil production flowlines are insulated by design to maintain the temperature as high as possible in the flowstream before arrival at the platform. In contrast gas–dominated systems cool much more rapidly compared to oil-dominated systems, requiring inhibitor injection to prevent hydrate formation, as discussed earlier.

To illustrate where hydrates might form in a process, consider the simplification of an offshore system shown in Figure 2.2. In Figure 2.2 the hydrocarbon flows from the reservoir up the well and through the Christmas tree or wellhead (which is comprised of many valves), typically through a manifold, into a flowline, which may be 30 to 100 miles in length, before it rises to a platform. The major tasks of the platform are fourfold: (1) separate the gas, oil, and water before disposing of the water; (2) compress the gas; (3) pump the oil to the beach; and (4) remove (dry) the water from the gas before inserting the gas into the export line, where it will flow to the beach.

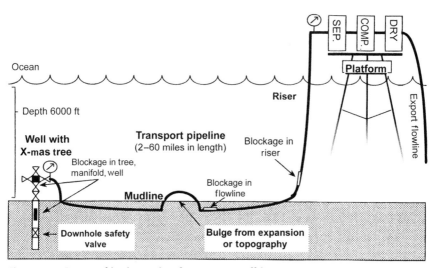

Figure 2.2 Points of hydrate plug formation in offshore system.

In the above system, production of high-energy-density oil is best (i.e., greater energy per unit volume); however, invariably gas and water are also produced, providing the components for hydrate plug formation. The SCSSV (the surface controlled subsurface safety valve in Figure 2.2) is placed at depth, so that the heat of the earth maintains a sufficient temperature above a hydrate stability condition at shut-in pressure and temperature to prevent hydrate formation. However, as indicated in the discussion of Figure 2.1, the flowline fluid rapidly cools to approach the temperature of the seafloor, which is usually within the hydrate stability region. Hydrates will not form in the export line due to the absence of water, unless there is a dehydrator malfunction.

Inhibitor is injected at the wellhead to prevent hydrate formation at any point above the SCSSV in Figure 2.2. Typical points of hydrate accumulation are (1) downstream of flowline water accumulations, such as a flowline low spot or at the riser; (2) when water has accumulated in the line or well during a shutdown with associated cooling; (3) beyond a restriction (e.g., a choke valve at the tree or fuel gas takeoff line); or (4) in the export line when a dehydrator failure has occurred. Hydrates may also form in well drilling mud systems and other places in the production system if the ingredients (water and gas) for hydrate formation are present at appropriate pressures and temperatures. Such occurrences are considered in Sloan and Koh (2008, pp. 19, 28) and works cited therein.

When the flow system is shut in, hydrate forms a thin film on the settled aqueous phases. Upon restart of flow, the produced fluid may disperse the resident water in the flowline, as water droplets that are gas saturated and immediately form an external hydrate film on the droplets. The accumulation of the hydrate-coated water droplets results in hydrate plug formation. Hydrate formation continues from an initial, thin hydrate film encapsulating water droplets that agglomerate to solidify the accumulated mass and form a plug. It has been shown that hydrate plugs may be as little as 4 volume% hydrate (Austvik, 1992), with the remainder encapsulated as liquid water, although the agglomerated, hydrate-encrusted droplets act as a hydrate plug.

2.2 HOW DO HYDRATE PLUGS FORM? FOUR CONCEPTUAL PICTURES

When oil and gas are produced into a flowline, they are invariably accompanied by water, so that three phases are commonly present: liquid hydrocarbon, aqueous liquid, and gas. Since a unified hydrate formation flow model does not exist, the problems will be described by breaking it into four end-member models, and presented in this section.

1. **Oil-Dominated Systems.** These systems have gas, oil, and water, but are dominated by the presence of oil, in which all of the water is emulsified as droplets in the oil phase, either due to oil surfactants or shear. Here the oil holdup would typically be 50% (volume) or greater.

2. **Gas-Dominated Systems.** Gas-dominated systems have small amounts of liquid hydrocarbon or aqueous liquid present. These systems are the few that have documented field data for hydrate blockage.

3. **Gas Condensate Systems.** These differ from oil-dominated systems in that they cannot disperse the water in the liquid hydrocarbon phase. Condensate systems are defined here to have water dissolved in the condensate, or suspended as droplets in the condensate due to high shear.

4. **High-Water-Cut (Volume) Systems.** When the water content is large (water holdup typically greater than 70% volume), such that water can no longer be totally emulsified in the oil phase, a separate continuous water phase exists. We limit these studies to conditions below the inversion point—those systems that have water droplets suspended in the oil, and oil droplets suspended in a separate water phase. Actual phase inversion has infrequently been observed in oil systems to date. In the maximum amount of water considered, there are two liquid phases, an oil phase with emulsified water droplets, and a water phase with emulsified oil droplets.

In the remainder of Section 2.2, the previous four end members are considered.

2.2.1 Hydrate Blockages in Oil-Dominated Systems

The following conceptual picture was generated by Turner (2005) with input from J. Abrahamson (University of Canterbury, Christchurch, NZ), for hydrate formation in an oil-dominated system with small (<50 vol%) water cuts, shown schematically in Figure 2.3.

In the conceptual drawing of Figure 2.3, four steps lead to hydrate plug formation along a flowline:

1. Water is dispersed in an oil-continuous phase emulsion as droplets, typically less than 50 μm diameter, due to oil chemistry and shear.

2. As the flowline enters the hydrate formation region in Figure 2.1, hydrate grows on the droplet rapidly (1 mm/3 sec [Freer, 2000]) at

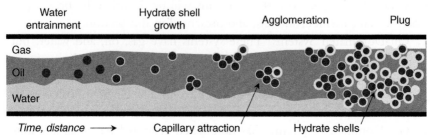

Figure 2.3 Conceptual picture of hydrate formation in oil-dominated system.

the oil–water interface, forming thin (5–30 μm thick) hydrate shells around the droplets, with the particle size unchanged (Taylor et al., 2007). This concept was discussed in the final section of Chapter 1.

3. Within each hydrate shell, shrinking-core droplets continue to grow, as a function of mass transfer of the guest and water through both the oil and the hydrate shell and heat transfer, dissipating the energy from hydrate formation. There may be free water within and between the droplets, which enables strong capillary attractive forces between the hydrated droplets. Water droplets can also wet hydrated droplets and/or be nucleated by hydrated droplets.

4. Hydrate-coated droplets will agglomerate to plug the pipeline, as shown at the right in Figure 2.3. This plug is initially mostly water, encapsulated within small hydrate crusts, although the plug acts like a solid and may continue to "anneal" to a more solid-like structure over time.

The above conceptual picture has evolved from a decade of both laboratory and flowloop studies (Boxall, 2009). The key to preventing hydrate plug formation is to prevent agglomeration by cold slurry flow, anti-agglomerants, or other techniques, such as in naturally inhibited oils (Camargo and Palermo, 2002). However, a key implicit assumption is the emulsification of all water in this oil-dominated model.

2.2.1.1 Rules of Thumb for Hydrate Formation in Oil-Dominated Systems

From the experiments used to generate the above conceptual picture, five rules of thumb arise for hydrate flow assurance in oil-dominated systems:

1. The formation of emulsions that keep the water/hydrates dispersed is generally a necessary step in resisting blockage formation (Turner, 2005).

2. There are two requirements to prevent hydrate plug formation in oil-dominated systems:
   - Low concentration of hydrate particles (<50 vol%) is needed; a higher particle concentration cannot be transported by the oil, and thus results in a plug.
   - Particle aggregation is prevented when the particles are oil-wet through oil chemistry, or by application of anti-agglomerant chemicals (Aspenes et al., 2008; Borgund et al., 2008; Buckley et al., 1998).

3. The closer the operating conditions are to the hydrate dissociation conditions, the stronger the interactions are between hydrate particles (Taylor et al., 2007), due to attractive capillary forces from a quasi-liquid layer at the particle surface.

4. The formation and dissociation of hydrates can cause coalescence of hydrate droplets that are dispersed in oil (Greaves, 2007; Høiland, et al., 2005). This coalesced phase can easily combine with gas molecules to form a blockage.

5. Like other deposits, freshly formed hydrates are more porous and malleable than are hydrates that have time to age and "anneal." The process (something akin to Ostwald ripening, which minimizes the surface area to volume to minimize the system free energy) causes a more dense crystal mass, which makes dissociation of the plug increasingly difficult (Davies et al., 2006).

2.2.1.2 A Model for Hydrate Formation in Oil-Dominated Flowlines

CSMHyK is a module for the OLGA® (SPT Inc.) transient, multiphase, flow simulator; CSMHyK has been under development since 2003. This model simulates hydrate plug formation in flowlines using the transient, multiphase flow simulator, OLGA®. Because detailed data from hydrate blockages in oil-dominated field flowlines do not exist, validation of the program to simulate hydrate formation in pipelines has been limited to experiments in flowloops for oil-dominated systems. These experiments have been used to verify and further develop the CSMHyK hydrate kinetics model.

The model predicts the rate of hydrate formation using a first-order rate expression proposed by Bishnoi et al., (1989), based on the thermal driving force and the rate based on Bishnoi's "intrinsic" rate constant for hydrate formation. Previously, CSMHyK was fit to the growth rate of hydrate in the ExxonMobil Friendswood flowloop by using a multiplier of 1/500 times Bishnoi's intrinsic rate constant k_B, (Davies, 2009).

The CSMHyK-OLGA® model (Boxall et al., 2008) assumes that water droplets fully convert to form solid hydrate particles. Nucleation has been adjusted to match the observed subcooling in the flowloop so that the initial growth of hydrates can be matched to the observed initial rate in the flowloop experiments from Texaco's Bellaire flowloop, Exxon Mobil's Friendswood flowloop, and the Marathon flowloop at the University of Tulsa. A nominal subcooling of these facilities has been observed on average to be ±6 °F prior to hydrate formation. Once formed, the model assumes that these particles remain dispersed in the oil phase. The change in relative viscosity of this phase is then found based on the Camargo and Palermo (2002) expression for steady-state slurry flow. An overview of the current CSMHyK module and its integration into OLGA® is shown in Figure 2.4.

Six different hydrate formation experiments using the University of Tulsa flowloop have been compared against simulations using the CSMHyK-OLGA® flowloop model. The fitted parameters of the data from one oil in the ExxonMobil flowloop were used for all of the simulations of four other oils in the University of Tulsa flowloop. The remarkable result was that the average value fit for one oil on the ExxonMobil flowloop satisfactorily predicted the University of Tulsa flowloop hydrate formation experiments on four oils, as shown in Figure 2.5.

A 10-km model flowline shown in Figure 2.6 was simulated by Montesi and Creek (2006), using the CSMHyK-OLGA® program. In Figure 2.6, the fluid in the pipeline was cooled by the 40 °F ocean water,

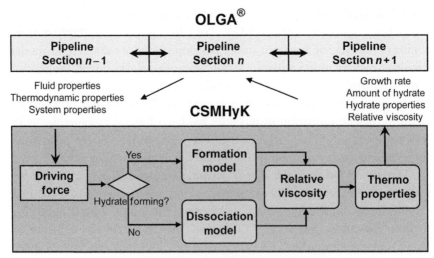

Figure 2.4 Integration of CSMHyK into the OLGA flow simulator.

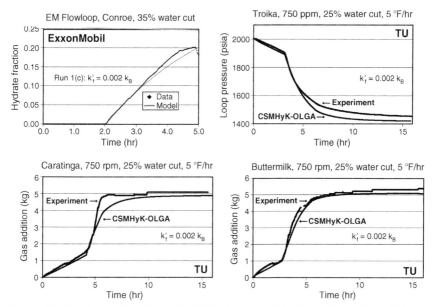

Figure 2.5 Fit of one oil on ExxonMobil flowloop predicts three other oils on University of Tulsa flowloop. There are some similarities of these flowloops from a dimensional and multiphase flow perspective.

below the hydrate equilibrium temperature (marked as "Eq T"). At about 6 °F below the equilibrium temperature, hydrates began to form due to the nucleation subcooling limit. The flowline temperature rose rapidly, on hydrate formation, to the hydrate equilibrium temperature due to the inability of this poorly insulated flowline to dissipate the heat of formation generated by the hydrate reaction. The hydrate formation heat generated by the reaction then maintained the flowline temperature at the equilibrium value, until the water that could be converted to hydrates at the flowline conditions was exhausted about 5 km from the source. Once there were insufficient reactants (water and hydrate formers) the reaction ceased, and the flowline temperature decreased to the sea bottom temperature, which was taken as 40 °F in this example.

This example suggests that the rate of hydrate formation is limited by the rate of heat removal from the pipeline with a heat transfer coefficient of 20 BTU/hr/ft^2/°F. The simulation results confirm that the CSMHyK-OLGA® model can be applied to flowline systems and to flow assurance design, in this case to heat transfer limited hydrate formation, and therefore blockage formation.

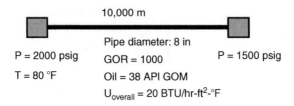

Alberto Montesi CVX

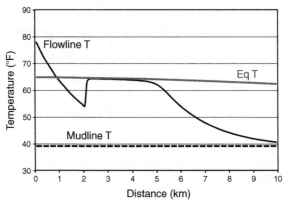

Figure 2.6 CSMHyK-OLGA® simulation of hydrate formation in a flowline (*Montesi and Creek, 2006.*)

2.2.2 Hydrate Formation in Gas-Condensate Systems

There are many cases wherein the mechanism reported in the earlier section is inadequate. For example, we have limited data for gas systems without significant amounts of liquid, but do have the following two field case studies, from the Tommeliten-Gamma field in the North Sea and the Werner-Bolley field in Wyoming. Those are the only two well-documented blockage studies in the field of which the authors are aware.

2.2.2.1 Case Study 1: Tommeliten-Gamma Field

In 1994, Statoil intentionally formed a series of eight hydrate plugs in a 6-in service line in the Tommeliten-Gamma North Sea gas condensate field. The layout of the field is shown in Figure 2.7. The simulated liquid holdup and flowline topology are shown in Figure 2.8.

It can be seen in Figure 2.8 that there was very little condensate holdup (<0.07) and water holdup (<0.01) in the steady-state flowing system.

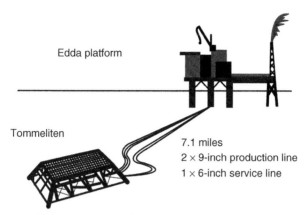

Figure 2.7 Tommeliten-Gamma lines feeding gas condensate into the Edda platform.

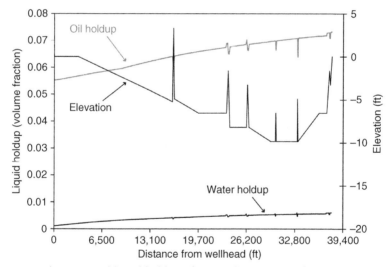

Figure 2.8 Elevation and liquid holdup of Tommeliten-Gamma field service line.

The reported spikes in the pipeline topology occurred when the line overpassed other pipelines, and resulted in coincidental spikes in the hydrocarbon liquid holdup and smaller spikes in water holdup. In the eight tests, hydrate plugs always formed downstream of the second and third spikes from the wellhead, where the pipeline entered the hydrate pressure and temperature stability region, and where there was sufficient water accumulation to form hydrates.

2.2.2.2 Case Study 2: Werner-Bolley Field Hydrate Formation

In the winter of 1997, Southwest Research Institute (SwRI) measured the hydrate plug formation and dissociation in a gas condensate pipeline at the Werner-Bolley field (Hatton and Kruka, 2002) located in southern Wyoming. The fluids, mainly composed of a gas phase (4 MM SCFD of gas, 100 BOPD of gas condensate, and 10 BWPD of water) flowed through the 3.3-mi long, 4-in diameter pipeline. The topographic profile of the Werner-Bolley pipeline is shown in Figure 2.9. Most of the flow pattern was stratified, but some flow slugs of water/condensate were observed. There were five data acquisition sites (bell holes), marked as solid square symbols in Figure 2.9, to measure the temperature and pressure of the system. The plugs were dissociated via one-sided dissociation and progressed rapidly down the flowline, with attendant safety precautions as suggested in Chapter 3. The plug length, velocity, and liquid-holdup were measured at Site 4 by gamma-ray densitometry. Table 2.1 gives the composition of Werner-Bolley fluid (Matthews et al., 2000).

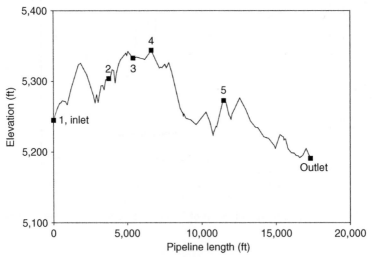

Figure 2.9 Elevation profile of Werner-Bolley field. Numbered squares represent points of P, T, and density (4 only) measurement.

Table 2.1 Composition of Werner-Bolley fluids

Composition	N_2	CO_2	C_1	C_2	C_3	$i\text{-}C_4$	$n\text{-}C_4$	$i\text{-}C_5$	$n\text{-}C_5$	C_6	C_{7+}	Sum
Mole %	0.43	1.51	78.2	13.6	3.61	0.39	0.69	0.25	0.21	0.24	0.87	100

Four trials of hydrate formation/dissociation were conducted, originally with the intention to determine if hydrate plug can be dissociated, via downstream, single-sided depressurization. Each test commenced by stopping methanol injection, pigging the pipeline with normal production, and then maintaining the production. Hydrate plug formation was indicated by pressure changes, which, for simplicity are indicated with only one of four pressure measurements in Figure 2.10. Typical hydrate blockage formation showed different behaviors with time.

Early stage: The pressure drop gradually increased from the upstream.
Middle stage: The cycles of pressure drop increase and collapse were observed in mainly the upstream section but sometimes in the downstream section.
Final stage: Developed hydrate blockage moved downstream and were stopped by restrictions.
Post-flow: Blockage was developed with limited or no flow.

The times required for hydrate blockage were between 56 and 143 hr. The pressure drops along the pipeline are summarized in Table 2.2. Times required for hydrate formation, when significant pressure change began,

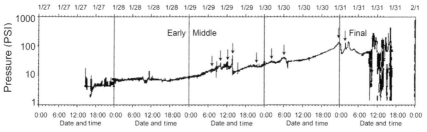

Figure 2.10 Pressure buildup and blockage in the Werner-Bolley line.

Table 2.2 Hydrate plugging time in Werner-Bolley field

Run	First observable pressure change (hr)	Time to first spike (hr)	Plugging time (hr)	Maximum pressure at the wellhead (psi)	Ground temperature (°F)
1	38	44	99	1030	44.6
2	36	48	98	1040	45.4
3	34	36	56	1120	45.4
4	72	110	143	>1200	43.8
Average	45	59	99	—	44.7

ranged from 34 to 72 hr, from the start of hydrate formation. It was suggested that hydrate accumulation caused spikes in the pressure curves of Figure 2.10 at the arrow indications. The final plugging time was neither a function of the maximum wellhead pressure nor of the ground temperature.

The dominant hydrate restrictions were inferred by pressure-drops for segments of the line length. The location of the dominant plug changed with time. Most of the largest pressure drops were observed between Sites 2 and 3 and Sites 4 and 5 in Figure 2.9.

These conceptual events simulate cases when hydrate formation is unexpected, such as due to inhibitor injection failure or dehydrator malfunction. In the Werner-Bolley field, hydrate formation was detected by pressure, temperature, and gamma-ray density sensors at five locations over the 3-mi line length, noting that it was difficult to distinguish between densities of dense-phases slugs, such as hydrates, ice, or water. Matthews et al. (2000) reported that in each case, hydrate formation occurred with two necessary conditions: (1) where there was sufficient water accumulation or drainage, just upstream of a mountain range, and (2) where the fluid temperature was below the hydrate formation point by around 6 °F. This is consistent with the flowloop findings of Matthews et al. (2000).

2.2.2.3 Hypothesized Mechanism for Gas-Dominated Systems

In both of the above gas-dominated field examples, it is apparent that hydrate formation relies on the conjunction of water accumulation and sufficient subcooling below the hydrate formation temperature. All experience points to no hydrate formation where there is only water accumulation (and hydrate formers) without subcooling, or subcooling without water present. Studies at the University of Tulsa flowloop (Dellecase and Volk, 2009) and the Colorado School of Mines laboratory apparatuses, indicate that when gas bubbles through a water pool at the hydrate pressure and temperature, hydrates are likely to form downstream of the accumulation upon creation of large amounts of surface area caused by gas release from the water. These blockages are likely due to accumulations of hydrate-encrusted gas bubbles. Figure 2.11 is a conceptual picture hypothesized for hydrate formation in gas-dominated systems, based on case studies such as the one described previously.

A second hypothesis for hydrate formation in a gas–condensate system states that it results from wall deposition. This hypothesis is discussed in

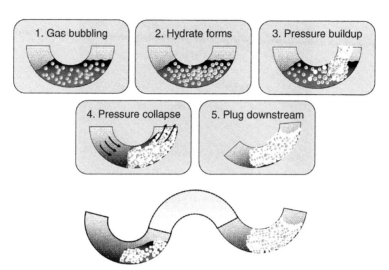

Figure 2.11 Hypothesis for hydrate formation from gas-dominated systems. (1) Gas bubbles through low-lying water accumulation. (2) When the subcooling is great enough (e.g., >6 °F), hydrate-encrusted gas bubbles form as the gas exits the water. (3) These hydrate-encrusted bubbles aggregate. (4) Collapse, forming (5) plug downstream of the original water accumulation.

detail in the following section for hydrate formation from a condensate, with evidence shown in the photograph in Figure 2.12.

2.2.3 Hydrate Blockages in Condensate Flowlines

Light, condensate-dominated systems differ from oil-dominated systems because stable water droplet emulsions do not occur, due to the lack of surface active components and hydrocarbon liquid viscosity. Documented, published field data are nonexistent for plug formation in condensate lines, although the general, flowline diameter-reducing (like arterio-stenosis) hydrate formation in a condensate flowline (Lingelem et al., 1994) has gained acceptance in the flow assurance community.

A conceptual picture and resulting rules-of-thumb for hydrate formation from a condensate were determined by Nicholas (2008). In work co-sponsored by Imperial Oil, Shell, and ConocoPhillips, Nicholas co-designed and performed measurements in a Westport Intertek gas condensate flowloop, shown in the simplified schematic in Figure 2.13. These measurements were in conjunction with a number of laboratory measurements of adhesive forces between condensate hydrates and pipe materials, showing that hydrates formed directly on the pipe wall will

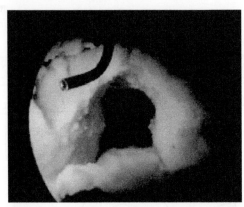

Figure 2.12 Gas-condensate hydrate formation narrowing the flowline channel via wall deposition. The top bent tube blows nitrogen against window to clear it for visibility. *(Courtesy of G. Hatton, Southwest Research Institute.)*

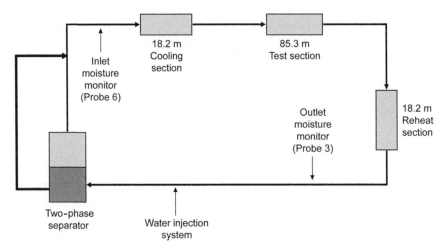

Figure 2.13 Simplified condensate flowloop schematic (Nicholas, 2008). Note that the moisture-monitored water concentration difference (inlet minus outlet) is due to solid deposits in the test section, which has pressure and temperature measurements at 40-ft (12.2-m) intervals. Hydrate formation temperatures were achieved at the exit of the cooling section.

adhere to the wall, whereas hydrates formed in the bulk will not adhere to the wall.

Some vital results from Nicholas (2008) are shown in Figure 2.14. In the figure, note that the difference between inlet and discharge condensate water composition is due to test section hydrate deposits, reflected in test

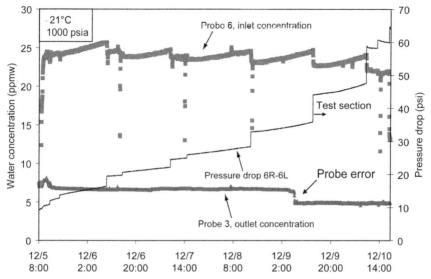

Figure 2.14 Difference between water inlet and outlet concentrations, and pressure drop buildup due to hydrate formation in a condensate loop. Note that step changes in pressure drop and concentrations are due to manual flow adjustments to maintain approximately constant flow rates. *(From Nicholas, 2008.)*

section pressure drop increases with time. Step increments in the figure reflect manual flow adjustments.

Figure 2.15 shows that, with only dissolved water, hydrates from condensate were deposited along the length of the flowloop, with thicker deposits at the beginning of the loop (40 ft), thinning to no deposits at the end of the test section (120 ft). Because hydrate wall deposits have very low thermal conductivity (20% that of ice), diminishing downstream temperatures (at 40, 80, and 120 ft) occur with length due to thinning hydrate deposits.

The condensate test loop work shown previously, together with other laboratory experiments (Nicholas, 2008), have enabled a conceptual diagram for the formation of hydrate from a condensate. A conceptual cartoon of how both a shrinking diameter and a resulting hydrate plug occur in a gas condensate line is shown in Figure 2.16.

A conceptual picture for hydrate formation in a condensate can be described with the following steps that correspond to the numbers in Figure 2.16.

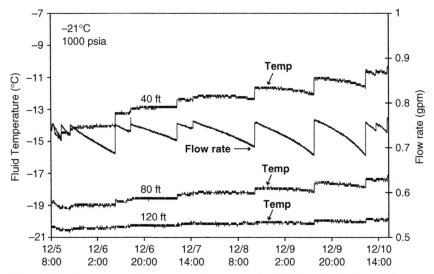

Figure 2.15 Temperature increases at 40-, 80-, and 120-ft lengths in the condensate flowloop test section due to hydrate wall deposits and propagation downstream. Flow rates show manual step adjustments to maintain approximately constant flow.

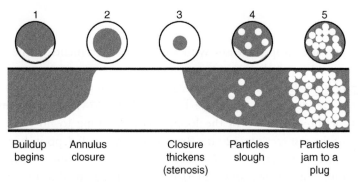

1	2	3	4	5
Buildup begins	Annulus closure	Closure thickens (stenosis)	Particles slough	Particles jam to a plug

Figure 2.16 One-way hydrate plugs form in a condensate flowline. In the figure, the darker shade represents condensate, while white represents hydrate.

1. Hydrates that begin to nucleate at the pipe surface will remain on the wall, dependent on water concentrations being higher than the hydrate stability limit in the condensate. This is usually caused by uninhibited water if the system is upstream of the platform, or dehydrator malfunction resulting in high water content in the gas export line.

   High (>7 ppm) concentrations of dissolved water provide a uniform, dispersed deposit along the flowline.

Free water results in a localized, early deposit as the flowline enters the hydrate stability pressure–temperature region, shown in Figure 2.1.

2. After hydrates nucleate at the wall at Point 1, they grow rapidly to encompass the entire circumference of the flowline.

3. As hydrates continue to grow, the effective diameter decreases analogous to arterio-stenosis in a blood vessel.

4. The large hydrate wall deposit builds until it is disturbed by some phenomenon, such as slug flow, density difference, harmonic resonance, and so on. At that point the deposit is no longer mechanically stable and sloughs from the pipe wall into the flow stream as hydrate particles.

5. The hydrate particles jam to a plug, preventing normal flow.

With hydrate deposition on flowline walls, the mechanism for condensate hydrate plug formation differs significantly from an oil-dominated system plug. In condensate systems, wall sloughing and particle jamming form a plug, while in oil-dominated systems, particle aggregation in the bulk causes plugging. Sloughing and jamming are active subjects of research in non-emulsifying fluids.

The previous conceptual picture has a number of important implications for operation of a gas condensate flowline. For example, if a platform dehydrator has malfunctioned to cause free excess water in the gas condensate, the export line should be shut in with immediate remediation. However, if the high water content does not result in a separate phase, but is in the form of dissolved water (yet above the hydrate equilibrium concentration), corrective action may then be taken by bringing the dehydrator to acceptable limits for dissolution of the hydrate wall deposit. Unfortunately, no rules of thumb have been validated for condensate systems, in contrast to the oil-dominated systems, for which five rules of thumb have been developed.

2.2.4 High-Water-Cut (Volume) Systems

In systems with high water cuts, such as occur in later field life, the water phase is not totally emulsified. A separate water phase occurs, as shown in the mechanism in Figure 2.17. Unfortunately, only hypothesized mechanisms such as Figure 2.17 can currently be provided.

Upon the continued addition of water, the water forms a separate phase. The inversion of the oil phase emulsion does not commonly occur, so that an external water phase remains.

The four conceptual pictures (oil-dispersed water, gas condensate, condensate only, and high [>50%] water cuts in Figure 2.17) represent the

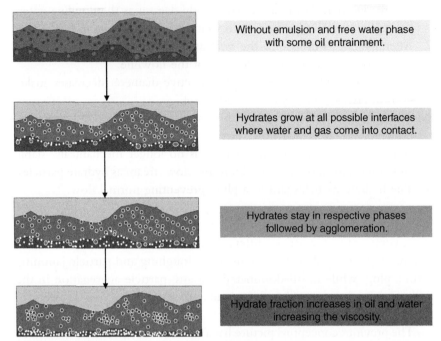

Without emulsion and free water phase with some oil entrainment.

Hydrates grow at all possible interfaces where water and gas come into contact.

Hydrates stay in respective phases followed by agglomeration.

Hydrate fraction increases in oil and water increasing the viscosity.

Figure 2.17 Hypothesis for hydrate formation in high-water-cut systems, when the water is not totally emulsified, but has an additional free water phase. *(From Joshi, 2008.)*

authors' best efforts to "connect the dots" of experimental evidence to form concepts that serve as foundations for later concepts of flow assurance and remediation.

2.3 RISK MANAGEMENT IN HYDRATE PLUG PREVENTION

In a growing number of flow assurance applications, hydrate risk management is more economical than avoidance of the hydrate formation region shown in Figure 2.1 (Sloan, 2005). In some cases the water cut, or amount, is so high that an uneconomical amount of inhibitor is required. For example, without recovery of methanol, the Canyon Express gas transport system would spend US$ 1million every 8 days for inhibitor injection (Cooley et al., 2003). As a second example, the Ormen-Lange

Norwegian field startup uses a large fraction of the world's annual production of monoethylene glycol to charge the inhibition system (Wilson, 2004).

In risk management, it is possible to allow hydrate particles to form in the gray area of Figure 2.1, and to prevent hydrate particle aggregation to a blockage, ensuring that the particles will flow dispersed in the oil phase. In order to move from avoidance to risk management, it is essential to quantify hydrate-formation time dependence, based on conceptual pictures such as those presented previously.

It is important to recognize that most of the quantitative data on hydrate kinetics was developed in the laboratory of Bishnoi and colleagues at the University of Calgary, particularly the pioneering studies of Englezos et al. (1987a, 1987b). The focus of this work was termed the "intrinsic" formation rate constant that was independent of mass and heat transfer limitations. Heat and mass transfer effects have traditionally plagued hydrate formation kinetics experiments, as typically the kinetic formation rate is much more rapid than either mass- or heat-controlled formation. In time-dependent experiments, the slowest phenomena tend to dominate the experimental observations. This was demonstrated in the CSMHyK-OLGA® simulations of Montesi and Creek (Figure 2.6), which showed that heat transfer could limit hydrate growth. A strict separation of all three effects (formation kinetics, mass transfer, and heat transfer) is required for acceptable modeling.

The next example illustrates how the previous concepts can be used for a new method of hydrate management.

2.3.1 Cold Stabilized Flow

The qualitative rules of thumb for oil-dominated systems in Section 2.2.1 have many flow assurance applications. These rules of thumb are quantified through application of CSMHyK-OLGA® (Boxall, 2009) predictions of the approximate rates of formation and fraction converted for a given flow scenario. It is worthwhile to consider cold flow as a potential application, as it has received much attention in both the industrial and academic communities.

The two main patented cold or stabilized flow concepts were developed by ExxonMobil shown in Figure 2.18, and by SINTEF Petroleum Research shown in Figure 2.19. The key principle in each process is to convert all free water droplets entirely to hydrate as rapidly as possible. With a minimal free water phase to provide capillary adhesion forces for hydrate particle aggregation, hydrate particles would not aggregate, but

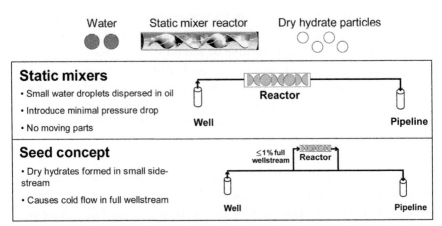

Figure 2.18 Cold flow in which particles are sheared and converted to dry hydrates that will flow with a condensate. *(From Talley et al., 2007.)*

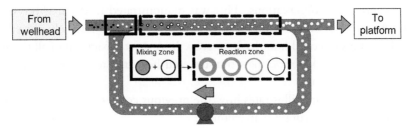

- Uses a preformed hydrate slurry to increase hydrate conversion rate

- Hypothesis: water coats hydrate particles, increasing surface area

- Rapid conversion rate of water minimizes agglomeration

Figure 2.19 SINTEF Petroleum Research cold flow concept. *(From Lund et al., 2004.)*

remain dispersed and flow with the oil phase. The typical analogy cited is that the flow would resemble "dry" snow, which is typically difficult to compact/aggregate into a snowball, compared with wet snow.

Details of how this technology might be implemented, and the relative volumetric flow rate of the incident and recycle streams, as well as other aspects of cold flow and other risk management applications arising from the above conceptual picture, are discussed in Chapter 5.

2.4 RELATIONSHIP OF CHAPTER TO SUBSEQUENT CONTENT

The conceptual pictures of how hydrate plugs form have direct consequences for how plugs are safely remediated in Chapters 3 and 4. Further, the conceptual pictures of plug formation are essential to understand the inhibition methods of Chapter 5, the certification of kinetic inhibitors in Chapter 6, and the evolution of operating procedures for hydrate control in Chapter 7.

REFERENCES

Aspenes, G., et al., 2008. Petroleum Hydrate Deposition Mechanisms: The Influence of Pipeline Wettability. In: Englezos, P. (Ed.), 6th International Conference on Gas Hydrates, Vancouver, Canada, July 6–10, 2008.

Austvik, T., 1992. Hydrate Formation and Behaviour in Pipes. D. Ing. Thesis, Norges Tekniske Høgskole Trondheim, Norway.

Bishnoi, P.R., Gupta, A.K., Englezos, P., Kalogerakis, N., 1989. Multiphase Equilibrium Flash Calculations for Systems Containing Gas Hydrates. *Fluid Phase Equilibria* 53, 97–104.

Borgund, A.E., et al., 2008. Critical Descriptors For Hydrate Properties of Oils Compositional Features. In: Englezos, P. (Ed.), 6th International Conference on Gas Hydrates, Vancouver, Canada, July 6–10.

Boxall, J., 2009. Hydrate Plug Formation from <50% water content water-in-oil emulsions. Ph.D. thesis, Colorado School of Mines.

Boxall, J., Davies, S., Koh, C., Sloan, E.D., 2008. Predicting When and Where Hydrate Plugs Form in Oil-Dominated Flowlines. In: Proc. Offshore Technology Conf., OTC 19514, Houston, TX, May.

Buckley, J.S., Liu, Y., Monsterieet, S., 1998. Mechanisms of Wetting Alteration by Crude Oils. In: SPE International Symposium on Oilfield Chemistry. Houston, Texas.

Camargo, R., Palermo, T., 2002. Rheological Properties of Hydrate Suspensions in an Asphaltenic Crude Oil. In: Mori, Y.H. (Ed.), Proc. Fourth Int. Conf. on Gas Hydrates, Yokohama, May 19–23, p. 880.

Cooley, C., Wallace, B.K., Gudimetla, R., 2003. Hydrate prevention and methanol distribution on Canyon Express. In: Proc 2003 SPE Annual Technical Conf and Exhib, SPE 84350, Denver Colorado, October.

Davies, S.R., Selim, M.S., Sloan, E.D., Bollavaram, P., Peters, D.J., 2006. Hydrate Plug Dissociation. *AIChE Journal* 52(2), 4016.

Davies, S.R., 2009. The Role of Transport Resistances in the Formation and Remediation of Hydrate Plugs. Ph.D. thesis, Colorado School of Mines.

Dellecase, E., Volk, M., 2009. Hydrate formation in the University of Tulsa flowloop, seminar at Colorado School of Mines, February 2.

Englezos, P., Kalogerakis, N., Dholabhai, P.D., Bishnoi, P.R., 1987a. Kinetics of Formation of Methane and Ethane Gas Hydrates. *Chem. Eng. Sci.* 42, 2647.

Englezos, P., Kalogerakis, N., Dholabhai, P.D., Bishnoi, P.R., 1987b. Kinetics of Gas Hydrates Formation from Mixtures of Methane and Ethane. *Chem. Eng. Sci.* 42, 2659.

Freer, E., 2000. Methane hydrate formation kinetics. M.S. thesis, Colorado School of Mines.

Greaves, D.P., 2007. The effects of hydrate formation and dissociation on high water content emulsions. M.S. thesis, Colorado School of Mines.

Hatton, G.J., Kruka, V.R., 2002. Hydrate Blockage Formation—Analysis of Werner Bolley Field Test Data. DeepStar CTR 5209–1.

Høiland, S., Borgund, A.E., Barth, T., Fotland, P., Askvik, K.M., 2005. Wettability of Freon hydrates in crude oil/brine emulsions: The effects of chemical additives. In: Austvik, T. (Ed.), Proc. 5th Int. Conf on Gas Hydrates, Trondheim, Norway, June 13–16, p. 1151.

Joshi, S., Presentation at DeepStar Meeting, Houston, TX, December 4, 2008.

Lingelem, M.N., Majeed, A.I., Stange, E., 1994. Industrial Experience in Evaluation of Hydrate Formation, Inhibition and Dissociation in Pipeline Design and Operation. In: Sloan, E.D., Happel, J., Hnatow, M.A. (Eds.), Proc. First Int. Conf on Natural Gas Hydrates. vol. 715. Annals of New York Academy of Sciences, p. 75.

Lund, A., Lysne, D., Larson, R., Hjarbo, K.W., 2004. Method and system for transporting a flow of fluid hydrocarbons containing water. Norwegian patent NO311,854, GB2,358640, Eurasian Patent 200100475.

Matthews, P.N., Notz, P.K., Widender, M.W., Prukop, G., 2000. Flow Loop Experiments Determine Hydrate Plugging Tendencies in the Field.

Montesi, A., Creek, J.L., 2006. OLGA Users' Group Meeting, Houston, Texas, November 27.

Nicholas, J.W., 2008. Hydrate deposition in water saturated liquid condensate pipelines. Ph.D. thesis, Colorado School of Mines.

Notz, P.K., 1994. Discussion of the Paper, "The Study of Separation of Nitrogen from Methane by Hydrate Formation Using a Novel Apparatus" In: Sloan, E.D., Happel, J., Hnatow, M.A. (Eds.), First Int. Conf. on Natural Gas Hydrates, Ann. N.Y. Acad. Sciences, vol. 715. p. 425.

Sloan, E.D., 2005. A Changing Hydrate Paradigm—From apprehension to avoidance to risk management. Fluid Phase Equilibria 228–229C, 67–74.

Sloan, E.D., Koh, C.A., 2008. Clathrate Hydrates of Natural Gases, third ed. Taylor and Francis, Boca Raton, FL.

Talley, L.D., Turner, D., Priedeman, D.J., 2007. Method of generating a non-plugging hydrate slurry. US Patent 60/782,449.

Taylor, C.J., Miller, K.T., Koh, C.A., Sloan, E.D., 2007. Macroscopic investigation of hydrate film growth at the hydrocarbon water interface. Chem. Eng. Sci. 62, 6524.

Turner, D., 2005. Clathrate hydrate formation in water-in-oil dispersions. Ph.D. thesis, Colorado School of Mines, Golden, CO.

Wilson, A., 2004. Personal Communication, Gas Processors Association Convention. San Antonio, TX. March 15.

CHAPTER THREE

Safety in Hydrate Plug Removal

Carolyn Koh and Jefferson Creek

Contents

More frequently than necessary the removal of hydrate plugs causes equipment to be damaged, and in the extreme, life is lost due to unsafe hydrate removal practices. This chapter provides case studies of hydrate safety incidents, and their underlying causes, so that safety problems can be avoided.

3.1 TWO SAFETY CASE STUDIES

3.1.1 Case Study 1: One-Sided Depressurization

Consider the plug removal incident depicted in Figure 3.1. In the winter, a hydrate plug formed in a buried, sour-gas (containing H_2S) flowline. Because the foreman had experience clearing such plugs, he was not unduly concerned about removal via depressurization. When he and his crew were called to a safety meeting at the home office, the foreman was content to let the plug sit for a few days. Unfortunately during the period of absence, the hydrate plug annealed and hardened, just as in Step 4 of the hydrate plug hypothesis in Figure 2.3.

Upon their return to the field, the foreman and an operator attempted to clear the hydrate plug by opening the downstream valve indicated in the inset drawing of Figure 3.1 to bleed the pressure in the downstream end from the wellhead. This pressure reduction caused the hydrate plug

Natural Gas Hydrates in Flow Assurance
ISBN 978-1-85617-945-4
DOI: 10.1016/B978-1-85617-945-4.00003-0

Figure 3.1 Hydrate plug removal incident from single-sided depressurization. Inset shows process flow diagram of line and valve. The hydrate plug was initially located at the lower left portion of the line. *(Courtesy of Chevron Canada Resources, 1992.)*

to partially dissociate, detach and launch from the wall. Personnel were standing near the line when the line failed, most probably from the pressure impact of a moving hydrate mass.

The fractured pipe erupted from the ground, and hit the truck, shown in the picture. By following the ground indentation trajectory (just to the left of the drawing insert in Figure 3.1) relative to the damaged, upraised truck hood, it appears that the pipe ruptured and erupted from the ground with a force sufficient to move the entire truck more than a yard (3 ft) to the side. A large piece of pipe struck the foreman and the operator summoned help. Due to the absence of safety air packs, the crew could not aid their injured colleague. An air ambulance was deployed; however, the foreman died before arrival at the hospital. No preexisting pipe defects were found. Two other very similar single-sided depressurization case studies are discussed in Sloan (2000, chapter 1).

3.1.1.1 The Cause and Effect of Hydrate Projectiles

Before proceeding to the next safety case study, consider the reasons for the above incident. Figure 3.2 shows a picture of a hydrate plug in a Petrobras slugcatcher. The agglomerated hydrate particles shown here

Figure 3.2 Hydrated particles accumulated from a flowline in a slugcatcher. *(Courtesy of A. Freitas, Petrobras.)*

were not the result of a plug forming in the flowline; rather the flowline was not sufficiently dewatered after a hydrostatic test on commissioning. When hydrate particles formed on line startup, they were pushed to the slugcatcher by a pig used to clear the line, resulting in the accumulation in the photograph.

If the pressure is lowered too much in an effort to clean the flowline blockage, simulations and experiments have shown that the hydrate will attempt to reach the equilibrium temperature corresponding to the low pressure value, which can result in ice formation and potentially lead to an ice-cased hydrate projectile (this likely occurred in Case Study 1, Figure 3.1).

Plug removal by heat is discussed in the next section. More typically, it is recommended for safety reasons that hydrate plugs should be dissociated and cleared via two-sided depressurization. The plugs are usually porous and permeable to gas, but not to liquids, which have higher liquid surface tension and viscosity. Upon depressurization the hydrate dissociation temperature can decrease significantly below that of the surrounding environment ($>39\ °F$), so heat flows into the pipe causing radial dissociation to commence at the pipe wall, as shown in the three separate time-interval experiments seen in Figure 3.3. If the pipe is insulated, the restricted radial heat flow can convert the hydrate plug to an ice plug.

Sloan and Koh (2008) discuss details of plug depressurization (Chapter 8), and provide a predictive computer program (Appendix B). For purposes of this book, it is sufficient to note the visual evidence for radial dissociation in the three separate plug dissociation experiments of

After 1 hr   After 2 hr

After 3 hr

Figure 3.3 Radial dissociation pictures of three hydrate plug experiments, in which the pipe was opened after 1, 2, and 3 hr of depressurization.

Figure 3.3. Because the plug dissociates radially, rather than longitudinally, the plug detaches at the pipe wall first.

Figure 3.4 shows the force difference across the plug $[(P_u - P_d)A_{cs}]$, where P_u and P_d are the upstream and downstream pressures, respectively, and A_{cs} is the flowline cross-sectional area. If the force difference across the plug is greater than the adhesion force to the pipe wall, such as during dissociation, then the plug will become a projectile in the pipeline. This can also happen when the dissociating plug shrinks from the wall allowing relatively easy detachment unless care is taken to have static liquid in the path to retard movement, as was done in the Werner Bolley DeepStar (Hatton and Kruka, 2002) experiments on one-sided hydrate plug dissociation.

In 1997, plug lengths in the DeepStar field studies in Wyoming (Hatton and Kruka, 2002) ranged between 25 and 200 ft, with measured velocities between 60 and 270 ft/sec. Because the hydrate density is that of ice (specific gravity of 0.9), it is much denser than the flowline

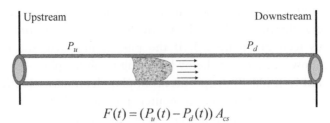

Upstream   Downstream

P_u P_d

$$F(t) = (P_u(t) - P_d(t)) A_{cs}$$

Figure 3.4 Force across hydrate plug due to pressure difference.

hydrocarbon density (typically a maximum of 0.6), and the plug will have much more momentum (mass times velocity) with little resistance to the flight of the detached hydrate mass.

As a result, the plug will flow down the line and significantly increase the pressure of the fluid in its path, perhaps to the pipeline rupture point, as shown in Figure 3.5b. This effect likely caused the pipeline failure in Case Study 1. A second cause of pipeline rupture occurs if the speeding hydrate plug encounters an obstruction such as a bend, indentation, or orifice, causing a sudden impact and subsequent rupture as shown in Figure 3.5a.

When the plug releases and is pushed downstream by the initial pressure gradient, the maximum pressure downstream will be from the initial plug thrust. The velocity will decrease due to friction, bends, liquids in the pipeline, and so on. If the plug is not slowed sufficiently in its initial trajectory, the velocity may oscillate with time, as shown in Figure 3.6 until the pressure equilibrates, when the plug velocity dampens to zero.

A more frequent occurrence than the damped multiple oscillations of Figure 3.6 was given by Xiao et al. (1998), who simulated the released plug velocity and condensate that retarded the plug in the Werner Bolley field, so that both the upstream and downstream pressure approached the equilibrium value without oscillation. The Werner Bolley plug departure velocity was captured by Xiao et al. (1998) with OLGA® simulations (estimated 300 ft/sec initial velocity).

3.1.1.2 Predicting Plug Projectile Effects

Figure 3.7 shows a prediction from CSMPlug (Sloan and Koh, 2008), the model for the downstream pressure in the general case of a plug release with single-sided depressurization. The plug mass in this study was 60 kg; pressure drop across the plug = 10 MPa; downstream pressure = 0.344 MPa; upstream pressure = 10.34 MPa; pipe diameter = 0.154 m; and the pipe length downstream of the plug = 4876 m, with the downstream:upstream volume ratio parameters given in Figure 3.7. Figure 3.7 shows the effect of changing the volume of gas upstream to downstream by varying the volume upstream of the plug. As the plug moves, gas is compressed at the downstream end of the plug, thus increasing the pressure. The pressure increases immediately, typically within 6 sec. As the ratio of volume downstream to upstream is increased, the maximum downstream pressure reduces dramatically as shown in Figure 3.7. This result is important for a pipeline operator, who can determine the maximum pressure relative to the pipe rupture pressure upon plug detachment for a given pressure drop.

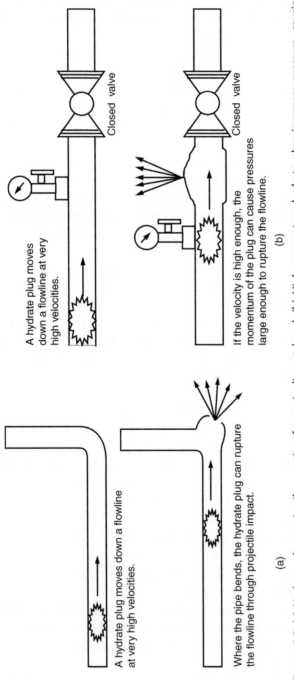

A hydrate plug moves down a flowline at very high velocities.

Where the pipe bends, the hydrate plug can rupture the flowline through projectile impact.

(a)

A hydrate plug moves down a flowline at very high velocities.

Closed valve

If the velocity is high enough, the momentum of the plug can cause pressures large enough to rupture the flowline.

Closed valve

(b)

Figure 3.5 (a) Hydrate plug projectile eruption from pipeline at bend. (b) High-momentum hydrate plug increases pressure, causing pipeline rupture. *(From Chevron Canada Resources, 1992.)*

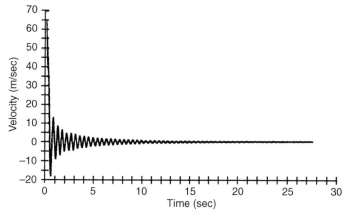

Figure 3.6 Plug velocity oscillation as a function of time, with dampening by friction and flowline liquids.

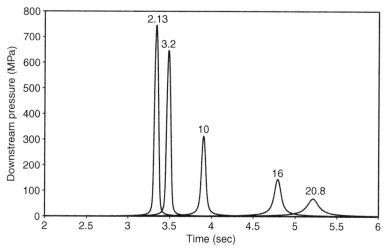

Figure 3.7 Effect of volume ratio (downstream to upstream of the plug, shown as parameters with each line) on downstream pressure maximum with time.

Based on Figure 3.7, the downstream pressure is below the pipe rupture pressure of 69 MPa (10,000 psia), when the ratio of volume downstream to upstream is around 21. Thus, the figure provides a safety estimation of the relative lengths downstream to upstream when the plug is released.

CSMPlug was used to obtain Figure 3.8 by changing the volume ratio to obtain an optimal pressure ratio for safe one-sided depressurization.

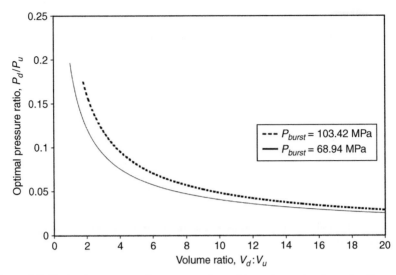

Figure 3.8 Optimal pressure ratio versus volume (downstream/upstream) ratio.

The graph is independent of plug mass and pipeline length. Figure 3.8 shows the optimal pressure ratio (downstream/upstream) as a function of volume ratio for two different cases when bursting pressure of the pipeline is 68.94 MPa (10,000 psia) and 103.42 MPa (15,000 psia). The bursting pressure of the pipeline depends on the pipe radius, pipeline thickness, material of pipe construction, and required safety margin. If a hydrate blockage is formed, the volume ratio can be estimated in the pipeline if both volumes and pressures can be determined upstream and downstream. For another way to calculate velocities, forces, and pressure, see Appendix A (pp. 82–86) of Chapter 4.

3.1.1.2.1 Example Calculation

Using Figure 3.8, determine the safe downstream pressure, with a pipeline bursting pressure of 103.42 MPa, and a volume ratio of 4, when the upstream pressure is 7 MPa (1015 psia). The optimal pressure ratio is 0.1, so the safe operating downstream pressure is calculated at approximately 0.7 MPa (101.5 psia). To be more explicit, a pressure drop of 6.3 MPa (914 psi) will not result in damage to this pipeline. The danger here in some systems with restricted heat flow may be that the dissociation temperature for sII hydrate is about 271 K (28 °F), leading potentially to hydrates being converted to ice. The ice formed may be more difficult to melt than the hydrates.

This example calculation and case study illustrate reasons why it is always better to (1) dissociate hydrate plugs slowly from both sides and

(2) assume that there are multiple plugs in a pipeline. With a slow, two-sided dissociation, if there is only one plug, there will be no pressure gradient across the plug when the plug dissociates radially, so no propulsion of the plug will occur.

3.1.1.3 The Effect of Multiple Plugs

When multiple plugs form in series in the pipeline, as shown in Figure 3.9, the situation becomes more problematic and even more precautions should be taken for plug dissociation. Multiple plugs may contain intermediate pressure(s), so it is prudent to decrease the pressure on the upstream and downstream side very slowly to maintain thermal and hydraulic control of the clearing process.

A second major type of hydrate safety problem results from heating a plug, as found in Case Study 2.

3.1.2 Case Study 2: Heating a Plug

In the case reported in Siberia in 2000, a pipefitter attempted to remove a hydrate plug by heating the exposed pipeline with a torch. Because the ends of the hydrate plug had not been accurately located, the pipefitter unknowingly heated the plug at the hydrate mid point. Gas pressure from the dissociated mid hydrate plug rose rapidly, confined by the plug ends, as shown in Figure 3.10. As a result of overpressure, the pipeline exploded. In the resulting fire, one man died and four others were severely injured.

The Siberian case study results from the second structural rule of thumb in Chapter 1, namely that if all of the hydrate cages are filled in sI or sII, the molecules are much closer together than they would be at ambient conditions. Hydrate concentrates the gas volume by as much as a factor of 180, relative to the gas volume at 273 K and 1 atmosphere.

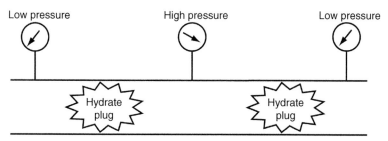

Figure 3.9 Safety hazard caused by multiple hydrate plugs that trap intermediate pressure. *(From Chevron Canada Resources, 1992.)*

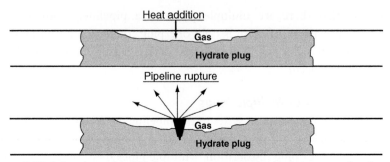

Figure 3.10 Safety hazards of high pressures trapped by hydrates upon heating the center of the hydrate plug. *(From Chevron Canada Resources, 1992.)*

In this case study, when the middle of the hydrate plug was dissociated by the torch heating the pipeline, every volume of dissociated hydrates released 180 volumes of gas, which were confined by the plug ends and caused the line to overpressure.

3.2 COMMON CIRCUMSTANCES OF PLUG FORMATION AND PLUG REMOVAL SAFETY

3.2.1 Common Circumstances of Plug Formation

In the previous two case studies, common equipment circumstances existed. Systems were out-of-service immediately prior to the incident, did not have hydrate or freeze protection, were pressurized while out-of-service, were being restarted, and had high differential pressures across plugs for short periods.

The Chevron Canada Resources Hydrate Handling Guidelines (Kent and Coolen, 1992) suggest that the danger of line failure due to hydrate plug(s) is more prevalent when:

• Long lengths of pressurized gas are trapped upstream of the plug.
• Low downstream pressures provide less cushion between a plug and restriction.
• Restrictions/bends exist downstream of the plug.

3.2.2 Plug Removal Safety Recommendations

The Canadian Association of Petroleum Producers Hydrate Guidelines (1994) suggests four safety concerns in dealing with hydrate blockages:

• Always assume multiple hydrate plugs; there may be pressure between the plugs.

- Attempting to move ice (hydrate) plugs can rupture pipes and vessels.
- Heating a submerged plug is not recommended due to the inability to precisely determine the location of the end of the plug as well as provide for gas expansion on plug heating. This is shown in Figure 3.10.
- Depressurizing a plug gradually from both ends is recommended.

In addition to the above recommendations, research has led to new safety recommendations over the last decade:

1. Locate plug(s) first and determine the volume upstream and downstream from the plug ends (e.g., using a mechanical device, gamma ray densitometer, or hoop strain gauge, reducing or cycling line pressure).
2. If one-sided dissociation of the plug is the only option, assess the jeopardy due to overpressure and plug projectiles, using a safety program such as CSMPlug (Sloan and Koh, 2008).
3. Treat plugs sooner rather than later to avoid the hydrate plugs from annealing/hardening over time (porosity and permeability decrease over time).
4. Electrical heating may be used when depressurization cannot be easily applied, such as when the liquid head on the hydrate plug is greater than the dissociation pressure, as in very deepwater or in mountainous terrain (Davies et al., 2006). However, the hazards associated with heating plugs need to be considered.

3.3 RELATIONSHIP OF CHAPTER TO SUBSEQUENT CONTENT

The safety considerations of this chapter are particularly emphasized in Chapters 4 and 7. Industrial standards for plug removal are found in Chapter 4, which extend the recommendations in this chapter with those used in practice. Chapter 7 is focused on industrial procedures for dealing with hydrates.

REFERENCES

Chevron Canada Resources, 1992. Hydrate Handling Guidelines: Safety and Loss Control Manual. Chevron Internal Report, July 23.

Davies, S.R., Selim, M.S., Sloan, E.D., Bollavaram, P., Peters, D.J., 2006. Hydrate Plug Dissociation. *AIChE Journal* 52 (12), 4016–4027.

Hatton, G.J., Kruka, V.R., 2002. Hydrate Blockage Formation—Analysis of Werner Bolley Field Test Data. DeepStar CTR 5209–1.

Kent, R.P., Coolen, M.E., 1992. Hydrates in Natural Gas Lines. Mobil Internal Report.

Sloan, E.D., 2000. Hydrate Engineering, Monograph 21, Society of Petroleum Engineers. Richardson, TX.

Sloan, E.D., Koh, C.A., 2008. *Clathrate Hydrates of Natural Gases*, third ed. Taylor and Francis, Boca Raton, FL.

Xiao, J.J., Shoup, G., Hatton, G., Kruka, V., 1998. Predicting Hydrate Plug Movement During Subsea Flowline Depressurization, Offshore Technology Conference, 8728.

CHAPTER FOUR

How Hydrate Plugs Are Remediated

Norm McMullen

Contents

Natural Gas Hydrates in Flow Assurance
ISBN 978-1-85617-945-4
DOI: 10.1016/B978-1-85617-945-4.00004-2

4.1 INTRODUCTION

Hydrate blockages can occur in many different and sometimes unlikely places. Circumstances surrounding the creation of blockages are often unusual and associated with human error. It is therefore very difficult to define absolute guidelines for every field situation. Regardless, many lessons have been learned using various approaches that do help narrow down the choices available for effective hydrate blockage removal.

It is important in all cases to think the problem through carefully before taking any action. In too many instances, operators will have taken steps beyond normal protocol in attempts to release blockages before calling in expert assistance, often exacerbating the situation. It is the investigators' challenge to determine exactly what is in the equipment and the circumstances surrounding the blockage formation.

In most cases, a team effort involving experts in this field will be required. Depending on the severity of the blockage and its impact on operations, pressure from management may be intense. It is important at times like these to remain calm and consider all options available before taking any action, as it is entirely possible to make matters even more challenging than they already are. There is much at stake, since remedial action involves the distinct possibility of risk to human life and assets.

Serious safety hazards and commercial impacts have already been discussed in Chapter 3, but the risks are again highlighted here. A sudden release of a plug with a high-pressure void behind it can have the same effect as an artillery projectile. Sudden changes in direction during its travel or a sudden restriction can rupture, distort, or dislocate equipment, resulting in loss of human life or damage to equipment involving expensive repair and downtime. It cannot be stressed enough that in many circumstances you will be dealing with a high-energy problem that is potentially unstable. Therefore, *all* suspected hydrate blockages should be treated as unstable until all safety issues have been neutralized.

It is very likely that the reader will turn to this chapter first if dealing with a suspected hydrate blockage. The chapter has been written with that in mind. The reader will learn the following from this chapter:

- How to determine whether the blockage is a hydrate plug.
- How to locate the plug and determine its size.
- Options available for successful blockage removal.
- Safety concerns when clearing hydrate plugs.
- Mechanical characteristics of plugs in different services (oil, gas, condensate) and the effects of mechanical characteristics on the remediation process.
- How to estimate the time required to clear a hydrate plug.
- How to deal with plugs in various types of equipment: onshore gas pipelines; onshore processing equipment, valves, and manifold; subsea wells; dry tree wells; subsea flowlines; and subsea manifolds, trees, and chokes.

4.2 SAFETY CONCERNS

Sudden releases of hydrate plugs have always been a concern in our industry and have caused fatalities and severe damage to equipment. The potential for the sudden release of significant masses of hydrate warrant careful management of potential trapped gas and high differential pressure in all systems. Calculations show that prudent pressure management and pressure diagnostic measures are required to prevent irreparable damage to equipment and potential release of significant volumes of hydrocarbons into the environment.

Depressurization has been the traditional and most effective means of dissociating blockages. Unfortunately, this potentially produces a high differential pressure across the plug, therefore producing a high driving force that could result in launching the plug at very high velocities, as discussed in Chapter 3. Hydrate plugs have been known to move at high velocities, resulting in line ruptures, equipment damage, and fatalities. Therefore, before initiating any depressurization strategy, it is important to determine whether a sudden release of any plug would result in damaging force in the path of the plug. Methods and tools have been developed for determining plug movement and what distance would be considered low risk during remediation (e.g. The chapter appendix or CSMPlug in Sloan and Koh, 2000).

Direct application of heat is also hazardous. In the past it has been generally considered inappropriate to directly apply heat externally to a zone containing a hydrate plug. This has been seriously debated in industry recently. Experiments have been constructed that employ judicious application of heat through direct electric heating. Regardless, unless the facility has been built around such a concept, it is unlikely that this method would be considered safe in most situations. Since 1 ft^3 of hydrate, which is an average value of 90% filled, will evolve about 164 ft^3 of vapor at standard conditions, pressure management would be imperative.

Human lives have been lost using both methods, and therefore require great care in their application.

Multiple blockages can and do create greater challenges. The possibility of the existence of multiple blockages should always be assumed. It is very likely that a pocket between any two blockages will remain at the pressure that existed during the formation of those blockages, which is likely to be the process that upset system conditions. Even during two-sided depressurization, there may be sufficient energy stored in the vapor space(s) between plug masses to launch a projectile at high velocity. Therefore, it is recommended that *all hydrate blockages be assumed to be multiple blockages* until proven otherwise. It is also recommended that the pressure in *any voids between those blockages be assumed to be at the highest upset pressure at the time of the blockage* until proven otherwise.

Pressure control is essential in all cases. It is not unusual for operators to immediately blow down a blocked system in an attempt to restart the system—perhaps hoping to "break" the plug. This can result in movement of fluids in the system and packing of what was initially a "soft" plug into a much more densely packed plug, and therefore a more difficult one to dissociate. Not only is it imperative to maintain control throughout the dissociation process, but also it is equally important to maintain pressure balance at process conditions across the plug(s) until a strategy has been forged.

4.3 BLOCKAGE IDENTIFICATION

The very first step in dealing with any blockage is to determine what the blockage is. Although it is usually obvious to those intimately familiar with the system, it is still very important to determine what is preventing flow in the system before taking any next step.

4.3.1 Determining Cause of Blockage

Although a statement of the obvious, it is nevertheless important to reexamine what was happening at the time that the blockage occurred. It is also important to consider what was done (or not done that should have been done) that could have led to blockage. Possibilities for blockage include:

- Mechanical (pigs, crushed pipe, construction detritus, etc.)
- Scale (slowest forming)
- Asphaltene blockage (slower forming)
- Wax blockage (slow forming)
- Composite (quickly forming gunk, chemically induced due to compatibility issues)
- Hydrate (hydrate formers needed, rapid formation, process upsets, inadequate treatment, etc.)

It is important to determine early on whether any pigging operations were underway, and whether there are mechanical objects within the system. Less obvious are road crossings where years of heavy equipment crossing the pipeline have ovalized the pipe, creating sufficient restriction to attract deposits and finally block the system.

The pressure buildup is more gradual with scale, asphaltenes, and paraffins, as deposition on the periphery of the pipe wall causes a gradual increase in line pressure drop.

Scale deposits that block systems are usually slow in developing. An exception may be found in new well completions whose chemistry has not been adequately addressed. The investigator still should rule this out as a possibility.

Asphaltene blockage is also a slow mechanism, requiring months to establish itself as an issue; often occurs in wells. More rapid accumulation can occur in separation equipment, but normally will have been discovered through daily sampling and other process upset indications, such as emulsion problems.

Wax blockages typically occur over a period of months, but can happen rapidly when accumulated in front of a poorly designed pig run. Wax deposits resulting in plugging will not occur in stagnant conditions, which means that wax plugging can be eliminated as a possibility in dead legs or future connections to equipment.

A more quickly forming blockage can occur when composites of precipitates resulting from chemical imbalances in the system that were not properly addressed, coagulate in production systems. Such blockages can appear to be hydrate plugs because of their rapid formation—especially when potential hydrate formers are also present. This sort of occurrence is most likely during startup of production systems when completion brines are still present in the production stream.

Hydrate plugs will usually form very quickly when conditions are right. Most hydrate blockages accumulate in the span of hours or a few days, but the final plugging event usually happens very rapidly, and often dramatically. Key indicators include dramatic fall-off of water receipts, dramatic downstream pressure reduction, and response to rapid shut-in and remedial action such as injection of methanol into a cold well following a restart abort sequence. Temperature, pressure, water, and hydrate gas-former content must be satisfied in order for a hydrate blockage to exist, and can be quickly included or eliminated from the list of possibilities on this basis alone.

4.4 LOCATING BLOCKAGE

It is rarely possible to precisely locate a plug, and for practical purposes, its exact location may not need to be known in order to address remediation. Still there are a number of ways one can determine plug location.

Use logic in a time when others may be emotional. Examine the circumstances leading to blockage. Look at process data available. Modern offshore systems are well instrumented and can provide valuable insight as to where high-pressure differentials existed and grew over time. Temperature and pressure data will provide key indications of where hydrates can form and where they cannot. This helps in establishing problem constraints.

During mechanical attempts to clear an unknown blockage, especially in wells (dry tree offshore), wireline or coiled tubing contact depth can give highly accurate location data of the plug. This occurs frequently during well work on mature assets, where wireline operations were partially or wholly responsible for creating the blockage.

Prediction by simulation along with process data (for model tuning) can often provide reasonably accurate location predictions, since gas void fraction will be part of the prediction. Steady-state simulation can often be used when a known event triggers the formation of hydrates. This would be the case, for example, when a dehydration plant goes offline, but wet export gas continues to be transported.

One can accomplish a simpler early prediction by injecting a known significant volume of thermodynamic inhibitor (such as glycol or methanol) and observing the pressure response of the system. Volumetric determination assumes the plug to be impermeable to the inhibitor and that the liquid holdup in the line is known (or negligible). Both may be incorrect since hydrate accumulations push substantial liquids ahead of the plug. This has proven to be an effective tool for gas production systems and gas export lines both onshore and offshore.

In recent years, gamma ray densitometry has played a larger role in both locating and quantifying plug characteristics. One must be able to access the location of the plug. A buried line cannot be accessed easily, but subsea lines can and are accessed routinely. It is impractical to launch a workboat and sample pipeline locations with such a tool unless preliminary location predictions have been made. Once the plug has been located, a great deal of important information can be extracted by the densitometer such as density distribution, multiple plugs, gas voids between multiple plugs, and probable liquid distribution.

Another innovative tool that has recently been used in the Gulf of Mexico is the hoop strain gauge. The tool is installed by a remotely operated vehicle (ROV) around the pipe. The gauge reports any change in hoop dimension. This tool is especially useful for determining if plugs are sealing, and therefore trapping gas volumes that might propel the plug at high velocity if dislodged during depressurization.

4.5 DETERMINING BLOCKAGE SIZE

Estimating plug size is necessary. Both length and mass are needed. The estimate is usually a judicious guess, so it will always need to be conservative.

Process data can be used not only to predict likely plug locations, but also can be used to predict plug size. Given the length of time that hydrate formation conditions have been present in a system and the likely passage of water into and out of the system, one can predict the volume of hydrate likely accumulated in the system. If one assumes that the accumulated water in the system is entirely converted to hydrate, an upper limit can be placed on the total hydrate in the system.

As mentioned above, gamma ray densitometry can give considerable insight as to the characteristics of the plug. It can also show boundaries of the plug or plugs, characteristics of the voids between and outside the plug or plugs, and a rough estimate of probable difficulty in permeating the plug with solvent. A densitometer can also be used to monitor progress during dissociation.

Other simple logic can also be applied. Information concerning the geometry of the system and probable location can be used effectively. For example, water is likely to pool at low points in the line. Well and flowline jumpers present excellent water traps for accumulating hydrates and subsequently plugs. Pipelines on an upward slope will tend to harbor a liquid inventory and present a water inventory for hydrate formation and ultimately plugging. It is likely in all of these instances that the plug size would be determined by the extent of the system's hydrodynamics and geometry.

It is also important to estimate the plugs' physical properties. They are necessary in assessing the dynamics of projectile risk and time to dissociate. If densitometer information is available, this task becomes easier. Accurate information is rarely available, so one has to assume the worst-case scenario. Necessary key parameters are plug density and shear strength. Plug density is needed to estimate time to total dissociation and the mass of the resulting projectile. Shear strength is needed to determine maximum safe differential pressure that can be applied to the plug before it will release.

Based on experience, the following are recommended for computational work:

Hydrate shear strength (Bondarev et al., 1996)	40 N/m^2	58 lb/in^2
Hydrate plug density	920 kg/m^3	57.4 lb/ft^3
Ice shear strength (Michel, 1978)	85 N/m^2	123 lb/in^2
Ice plug density	917 kg/m^3	57.2 lb/ft^2

Ice properties have been given for comparative purposes, and may be necessary for use should the plug be located in an environment where the plug will convert to ice. Note that Borthne et al., (1996) cited a much lower shear strength of 5.6 N/cm^2 (8.1 lb/in^2), but the method employed is suspect and does not match field observations.

4.6 BLOCKAGE REMOVAL OPTIONS

There are essentially four ways to remove hydrate blockages: pressure reduction, chemical application, mechanical removal, and thermal application. Each has its merits and risks. These options are discussed below.

4.6.1 Pressure

Dissociation by reducing pressure below the dissociation pressure at ambient temperature is most widely used in industry, where chemicals cannot be easily delivered to the plug location. The concept is straightforward. For any situation, compute the dissociation pressure for a given ambient temperature, and reduce the pressure evenly, if possible, on the plug. The lower the pressure achievable, the more rapidly the plug will melt. The goal of this method ultimately is to completely remove the plug. Once pressure communication has been established across the plug(s), it may then be possible to flood the system with thermodynamic inhibitor to accelerate the dissociation process, and stabilize the resultant mixture in preparation for cleanup operations.

Unfortunately, subsea conditions (~4 °C) call for pressures around 200 to 130 lb/in^2, conditions that may not easily be reached because of hydrostatic head in subsea systems. Arctic-like systems may initiate as hydrate plugs and subsequently convert to ice, making pressure reduction inapplicable.

Care must be taken when depressurizing. One-sided depressurization can work, but has the added risk of launching a partially dissociated plug toward the low-pressure end of the system. Prior to attempting to depressurize, it is essential that the risk of plug release be thoroughly assessed. This can occur whenever a plug or plugs at pressures greater than the required dissociation pressure trap a closed volume of vapor.

Two-sided depressurization is usually preferred as it reduces the total energy in the system and increases the speed at which the plug may be removed.

Finite difference simulators have been developed for estimating plug velocity and travel distance (Sloan and Koh, Appendix B, CD program CSMPlug). Given estimated plug size and density, and position, size, and pressure of voids, velocity versus distance can be computed for use in mechanical load computations. Such a simulator is not necessary, however, since an analytical solution exists (see the Appendix to this chapter). Further, it is not necessary to have a solution in the form of distance versus time, since the key information needed is plug velocity versus distance. It is important to know how fast the plug is going when it enters equipment, elbows, tees, valves, and so on. Clearly, plugs immediately adjacent to equipment and pipe fittings present the highest risk of damage. Presence of liquids filling the system adjacent to the downstream side of the plug can significantly attenuate plug velocity and should be considered in any analysis.

4.6.2 Chemical

It is difficult to get an inhibitor such as methanol or ethylene glycol next to a plug in a pipeline without an access method near the plug. Although plugs have been proven to be very porous and permeable, in gas systems a substantial gas volume between the plug and injection points (platform or wellhead) hinders contact, particularly when the line cannot be depressurized to encourage gas flow through the plug.

Inhibitors must therefore displace other line fluids through density differences to reach plugs. Usually opportunity is greatest when the plug is close to production facilities or subsea manifolds or trees. In pipelines with large variations in elevation, it is unlikely that an inhibitor will reach a plug without flow. Still, standard practice is to inject inhibitor from both sides of a plug, in an attempt to get the inhibitor next to a plug. Sometimes the increased density of heavy brines can provide a driving force to the hydrate plug face.

Methanol or glycol injection is normally attempted first in a line. Density differences act as a driving force to get the inhibitor to the face of the plug, resulting in glycol being used more than methanol.

Recent developments have shown that certain gases may also act as a solvent. Nitrogen and helium, for example, can easily permeate and dissociate hydrate plugs. Since hydrate plugs are typically gas permeable, this method shows great promise, but has yet to be field proven.

In all cases, chemicals that dissociate plugs are diluted by water and vapor released in the dissociation process, so they necessarily require continuous refreshing throughout the duration of the process.

Because many subsea systems and wells are now using low-dosage hydrate inhibitors (LDHIs), it is important to note that such chemicals may or may not have much value as solvents, depending on the carrier fluid used. Some products may in fact use methanol as a carrier, but would be very expensive (at current market) to deploy in large volumes.

Careful consideration must be taken in deploying these chemicals for the purpose of dissociating hydrates. The most likely application for LDHIs with appropriate carrier fluids would be in subsea wells where it is difficult if not impossible to gain access in any other fashion.

4.6.3 Mechanical

Coiled tubing has been used effectively where access is possible. This is especially true for dry tree facilities. The well can be entered using standard lubricator designs with coiled tubing. The tubing is extended down the well until the plug is tagged. Pressure balance is maintained on either side of the plug, preventing sudden movement. Either methanol or hot water is jetted against the plug face, eroding and dissociating the plug.

Hot water has been found to be very effective where heat transfer considerations prevent reforming of hydrates until the well fluids can be stabilized following removal of the plug and solids. The advantage of hot water is safety concerns in handling fluids via temporary hoses on the facility. Methanol is highly volatile and requires special handling, procedures, personal protective equipment, and so on.

Other devices, usually attached to the end of a coiled tubing assembly, have been proposed, but no successful operations using such a device have been documented. These include tractor pig designs or cup-shaped designs that help pull the coiled tubing farther into the system than it can be pushed. Drag forces limit the capability of this system. The more turns the tubing must navigate, the more limited the reach. Currently, the record distance achieved is about 5 to 7 mi. Clearly, a mechanical auger could speed up the process of clearing the plug and dissociating the hydrate solid phase.

A mechanism that extends the reach of coiled tubing for clearing wax plugs has been successfully deployed in the Gulf of Mexico. This system uses fluids pressure in the annulus between the coiled tubing and the production pipe to pull the coiled tubing into the pipe. At some time in the future it is anticipated that this system will be used to clear hydrate blockages.

Note that although workover drilling rigs have been used routinely for access to wells for remediation, drill ships have been used and proposed for pipeline/flowline and equipment intervention. This option requires extensive engineering and preparation time to execute. Clearly, this is an exceedingly expensive option for deepwater intervention but ultimately may be the only rational choice.

4.6.4 Thermal

As subsea flowline distance and water depth increase, heating is becoming more popular. The basic concept of the thermal approach is to increase the temperature of the hydrate plug above the equilibrium point (see Figure 4.1).

As temperature is increased above the equilibrium conditions, gas is released from the melting hydrate plug. If the gas can easily escape, then the pressure near the hydrate plug will not significantly increase. Note that in order for the gas to have a free path to escape, the entire length of

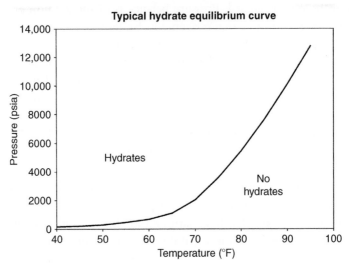

Figure 4.1 Typical hydrate equilibrium curve showing increasing temperature.

the hydrate plug must be at the same temperature. If the entire length of the hydrate plug is not at the same temperature, gas may be trapped, creating localized high pressures.

When that happens, the pressure near the hydrate plug will increase until the hydrate equilibrium pressure is reached. Hydrates will then start reforming. If the temperature of the hydrate plug was raised to 85 °F in this example, and there was no free path available for gas to escape, pressures near the hydrate plug could reach ~8000 psia. It is therefore imperative for any thermal method to have accurate temperature control.

Thermal methods to remediate hydrate plugs have been used intermittently in the field with no problem. However, there have been cases where pipes have burst and fatalities have occurred due to improper procedure. This is a safe and effective method of removing a hydrate plug if and only if the proper procedures are taken. Some of the following more common thermal methods typically are contemplated during project design.

4.6.4.1 Heated Bundle

A heated bundle consists of a pipe in which production fluids are flowed through the inner pipe and heated fluid is flowed through the outer pipe. A heated bundle is currently being used to heat the Gulf of Mexico King subsea multiphase flowlines. In this case, the tieback is looped so that heated fluid flows countercurrent outbound from the host, and co-current inbound to the host. No hydrate plugs have been encountered in the King lines. However, if a plug were to form, circulating water down the bundle would be considered a safe remediation technique. The heated bundle is considered a safe remediation method because the hydrate plug would be melted along its entire length and the heated fluid temperature can be controlled at the host.

4.6.4.2 Electrical Heating

Similar to a heated bundle, electrical heating consists of heating the external surface of the production flowline. However, instead of using a temperature-controlled medium, a thermal blanket, which applies constant heat flux, is applied to the pipeline. This method has been used for plugs in onshore pipelines in the Arctic. Nakika's North production system has an installed electric heating ready system to remediate plugs. The Nakika system was designed by Shell and British Petroleum is the operator.

In order for this method to be safe, it is essential that the thermal input be applied uniformly over the entire length of the hydrate plug. If the location and length of the hydrate are not known, the only safe option is to heat the entire line (ensuring that the entire hydrate plug is heated). A thorough analysis must be done to ensure that the temperature of the hydrate can be controlled within ±5 °F along the entire heating system.

The use of electrical heating to remediate hydrate plugs is controversial. In publications of limited distribution, Shell has argued that it is a safe practice. In contrast, Statoil has asserted opposition to the use of electrical heating to remediate plugs due to safety concerns. Others use inductive heating only as a last resort, and situations have been reported in which such usage nearly overpressurized the line. The high–pressure situation was created because the flowline was plugged at one end and blocked in by a valve at the other.

4.6.4.3 Heating Tent

Heat tents are commonly used for hydrate plugs in onshore Arctic pipelines that are above ground. The preferred method is to install blankets around the pipe and use a diesel fired heater to blow hot air across the suspected plug area.

4.6.4.4 Mud or Fluid Circulation

A common plug remediation procedure for wells is to circulate warm drilling mud (or some other fluid) down the annulus of the well.

The hydrate plug will melt radially from upstream to downstream. This method is similar to that of a heated bundle.

4.6.4.5 External Heat Tracing

This method employs hot water as a means to provide heat to a hydrate plug. Atlantis has adopted this method to remediate potential hydrate plugs in subsea equipment (jumper, manifold header, PipeLine End Termination (PLET), or tree). The hot water is circulated in an external tubing loop, and bonded and strapped onto the production pipeline, under the insulation. The loop is oriented axially along the full length of the affected production equipment such that inlet and outlet are adjacent to each other.

A remotely operated vehicle (ROV) connects via hot stabs to the tubing to allow circulation of the hot water. The ROV will be equipped with the pumping and heating skid that provides the friction-generated hot water. As the heat penetrates through the pipe, the local interior of the hydrate plug warms; a channel is dissociated along the plug, allowing pressure communication past the plug, thereby allowing flushing with thermodynamic inhibitor to complete the process. Because the remainder of the plug surface remains intact and adheres to the interior surface of the pipe, risk of plug movement is very low.

This method is a more desirable form of external heating because a communication channel would be established while the remaining plug would be intact and its structural integrity retained.

4.6.4.6 Guiding Principles for Thermal Remediation

Heating hydrate plugs involves additional considerations and thought. In many cases, the method employed should have been contemplated and included in the project design from the outset. Post-installation field deployment of most methods subsea is difficult and/or impractical. When using heat:

- The entire plug should be heated uniformly (this requires that the plug length and locations be known before remediation).
- Temperature should be controlled to within $\pm 5\,°F$.
- Pipe wall temperatures corresponding to dissociation pressures greater than maximum allowable pipe pressure may approach pipe bursting pressure.
- Heating with a point source (i.e., welding torch) is unacceptable.
- After a remedial action is taken to remove a plug, patience is required to observe the results over a long period of time.

4.7 REMOVAL STRATEGIES

Although many situations are similar, every hydrate blockage has its own unique set of circumstances. Consequently, every challenge requires attention to all details. However, in the early stages of assessment, it is helpful to have a template for guidance as to what steps to take and what options to consider.

This section discusses basic approaches and considerations for pipelines and flowlines, wells, risers, and process equipment. In all cases, it is important to recognize the impact and commercial implications of the blockage and act accordingly. If a specialist is available in the organization, the specialist should be promptly retained. This may be the case for large super majors, but smaller independents may not have such experts on hand. It is then advisable to retain an experienced consultant who specializes in hydrate remediation as soon as practical—even if it is later determined that there is no need. The reason for doing this is that these specialists are in short supply and typically are overbooked. The sooner scarce resources are retained, the better. Other specialists may also need to be booked as soon as the need is identified.

Valuable time can be lost brainstorming solutions that are already known to be impractical or of little use in the current circumstance. A number of computations may be necessary and an expert in the field can significantly shorten the time required to identify the solution, determine what needs to be done and when, and begin building high-level procedures to address the problem.

Additionally, if the facility affected has been designed for thermal remediation, the option is clearly to employ the planned method unless there are extenuating circumstances to the contrary. For completeness both considerations are repeated for each category below.

4.7.1 Pipelines/Flowlines Strategy

If the plug is within a safe calculated distance of the facilities, mechanical or chemical methods should be used in light of risks of safety and equipment damage. If the plug is more than the safe calculated distance from the facilities, depressurization is typically used to remove the plug. However, depressurization can take weeks to remove a plug. If quick removal is desired, mechanical methods, if available, could be another alternative.

4.7.1.1 Recommended Order of Consideration

1. Consult a specialist.
2. If the thermal method is included in the design, use the thermal method.
3. If the plug is closer than a safe distance from the facilities and a mechanical method is available, then use the mechanical method. Otherwise, use the chemical method.
4. If it is possible to depressurize the system below the hydrate equilibrium pressure, then use the pressure method.
5. If a mechanical device is available and the device can reach the plug, then use the mechanical method.
6. Nonstandard engineering solution is required.

4.7.1.2 Detailed Discussion of Pipelines/Flowlines Strategy

4.7.1.2.1 Pressure Method

Options
- Two-sided depressurization
- One-sided depressurization

Comments
- Do not increase pressure in the line. This will most likely make matters worse, making the plug harder and more compact, leaving a more difficult plug to remediate.
- After depressurization, the pressure at the hydrate plug should be less than the hydrate formation pressure at ambient temperature for this to be a viable option. Note that this pressure depends on water salinity.
- One-sided depressurization can be done if risks and concerns are considered.
- This option could take a considerable amount of time for the plug to be removed.
- After the plug is dissociated, ensure that proper amounts of chemicals are applied to prevent plug from reforming.

Recommendations
- Depressurize from both topsides and at the manifold. The lower the pressure, the faster the hydrate plug will melt.
- If two-sided depressurization is not possible, depressurize from topsides only if hydrate plug is more than a safe calculated distance from facility. If the plug is within a safe calculated distance from facility, attempt depressurization from manifold. It is essential to monitor pressure during the entire process. Jumps in pressure indicate plug movement.

Possible Risks and Concerns

- If the plug is near the platform or even in the riser, one- and two-sided depressurization could cause the plug to move at high velocities toward the platform, creating risks of rupturing the line, equipment damage at the facilities, and/or safety. Mechanical methods should be considered in this case.

4.7.1.2.2 Chemical Management

Options

- Methanol, ethanol, or glycol

Comments

- Enough chemical must be injected to fill the entire line to the plug. Therefore, if the plug is a great distance from the facility or from an injection point, there may not be sufficient chemical supply.
- This option could take a considerable amount of time for the plug to be removed.
- After the plug is melted, ensure that proper amounts of chemicals are applied to prevent the plug from reforming.

Recommendations

- Inject chemical into the line such that it fills the entire line to the hydrate plug. Preparation should be made to bleed off fluids already in the line to prevent high-pressure build up.

Possible Risks and Concerns

- Health, safety, and environment (HSE) risks associated with chemicals.

4.7.1.2.3 Mechanical Method

Options

- Coiled tubing

Comments

- Should only be considered if device can reach the entire length of the plug (typical coiled tubing reach is ~10,000 ft from the host).
- After the plug is dissociated, ensure that proper amounts of chemicals are applied to prevent the plug from reforming.

Recommendations

- Insert coiled tubing into line and jet brine or chemical (preferably hot) onto the plug. Calculations must be made on the minimum amount of brine or chemical to inject to ensure that once the hydrate is melted, the remaining fluids in the line are fully protected from hydrate formation.

Possible Risks and Concerns
- HSE risks associated with chemicals (if used).

4.7.1.2.4 Thermal Method
Options
- Electrical heating
- Heated bundle
- External hot water tubing

Comments
- These methods are considered during project design. If they are not currently implemented, other methods must be considered.
- Ensure that the location and length of the entire hydrate plug is known.
- Risks associated with electrical heating need to be assessed.
- After the plug is melted, ensure that proper amounts of chemicals are applied to prevent the plug from reforming.

Recommendations
- For any option, careful consideration of pipe wall temperature is essential for safe hydrate remediation. The outer pipe wall temperature corresponding to dissociation pressure that exceeds maximum pipe allowable working pressure should not be exceeded. Once the plug starts to melt, indicated by pressure increase, temperature can be ramped up.
- The heating medium must span across the entire hydrate plug.
- Uniform heating of the pipe wall is a must.
- Ensure that released gas has a free path to escape.

Possible Risks and Concerns
- If the pipe wall temperature is initially greater than that which corresponds to dissociation pressures greater than maximum allowable pressure, local pressures due to released gas from hydrate melting may exceed bursting pressure of the pipe.
- If the entire plug is not heated uniformly, released gas may not be able to escape, creating high local pressures, possibly exceeding bursting pressure of the pipe.

4.7.2 Wells Strategy
Mechanical methods of removing a hydrate plug (such as coiled tubing or wireline) are typically the most effective for wells in terms of safety and downtime. If mechanical methods are not available, chemical injection

and depressurization, which require a considerable amount of downtime, are the only alternatives. An exception to the above will be where thermal management systems are in place to deal with plugs.

4.7.2.1 Recommended Order of Consideration

1. Consult a specialist.
2. If the thermal method is included in the design, use the thermal method.
3. If a mechanical method is available, and the device can reach the plug, then use the mechanical method.
4. If warm fluids can be circulated down the annulus, then use the thermal method.
5. If there are chemical injection access points, then use the chemical method.
6. If it is possible to depressurize the system below the hydrate equilibrium pressure, then use the pressure method.
7. Nonstandard engineering solution is required.

4.7.2.2 Detailed Discussion of Well Strategy
4.7.2.2.1 Pressure Method
Options
• One-sided depressurization (from wellhead)
Comments
• Do not increase pressure in the well system. This will most likely make matters worse, making the plug harder and more compact, leaving a more difficult plug to remediate.
• After depressurization, the pressure at the hydrate plug should be less than the hydrate formation pressure at ambient temperature for this to be a viable option. Note that this pressure depends on water salinity. Also, note that ambient temperature will vary with depth and time.
• Keep a fluid cushion downstream of the plug at all times.
• After the plug is dissociated, ensure that proper amounts of chemicals are applied to prevent the plug from reforming.
Recommendations
• Decrease pressure at wellhead slowly.
Possible Risks and Concerns
• If a fluid cushion between the hydrate plug and wellhead is not kept, depressurization could cause the hydrate plug to travel at high velocities toward the wellhead, creating safety risks and risk of rupturing the tubing, riser, or flowline, or damage to equipment.

4.7.2.2.2 Chemical Method
Options
- Methanol, ethanol, or glycol
- Injection at the wellhead
- Injection at an access port (such as above the surface-controlled subsurface safety valve [SCSSV])

Comments
- Due to density differences, injecting glycol is preferred (as opposed to methanol or ethanol) at access ports above the hydrate plug to enhance the chances that the chemical reaches the plug.
- For similar reasons, methanol or ethanol should be injected at access points below the plug.
- Keep a fluid cushion downstream of the plug at all times.
- After the plug is dissociated, ensure that proper amounts of chemicals are applied to prevent the plug from reforming.

Recommendations
- Inject methanol or ethanol at SCSSV.
- Inject glycol at tree.

Possible Risks and Concerns
- HSE risks associated with chemicals.

4.7.2.2.3 Mechanical Method
Options
- Coiled tubing
- Heated wireline broach

Comments
- Keep a fluid cushion downstream of the plug at all times.
- Minimize pressure differential across the plug during remediation.
- After the plug is dissociated, ensure that proper amounts of chemicals are applied to prevent plug from reforming.

Recommendations
- Use coiled tubing to jet warm brine, mud, or chemical on the hydrate plug.

Possible Risks and Concerns
- HSE risks associated with chemicals.

4.7.2.2.4 Thermal Method
Options
- Mud/brine circulation

Comments
- Keep a fluid cushion downstream of the plug at all times.
- Minimize pressure differential across the plug during dissociation.
- After the plug is melted, ensure that proper amounts of chemicals are applied to prevent the plug from reforming.

Recommendations
- Inject warm drilling mud or brine down the annulus to heat the outer wall of hydrate plug.
- Once pressure communication across the plug is achieved, pump chemicals to further dissociate the plug.

Possible Risks and Concerns
- If a fluid cushion between the hydrate plug and wellhead is not kept, depressurization could cause hydrate plug to travel at high velocities toward the wellhead, creating safety risks and risk of rupturing the tubing, riser, or flowline, or damage to equipment.

4.7.3 Risers Strategy

Previous strategies given for wells are sufficient and more applicable to hydrate plugs encountered in a dry-tree or drilling riser. This section is dedicated only to plug removal options for a riser connecting a subsea flowline to the host. Mechanical methods of removing a hydrate plug (i.e., coiled tubing) are typically the most effective in terms of safety and downtime. If mechanical methods are not available, depressurization and chemical injection, which require a considerable amount of downtime, are the only alternatives. An exception to this would be where thermal management systems are in place to deal with plugs.

Also note that free standing risers that are then connected to a floating production and storage and offloading (FPSO) via flexible hoses would require a nonstandard engineering solution.

4.7.3.1 Recommended Order of Consideration

1. If this is a dry tree or drilling riser, refer to wells strategy above.
2. Consult a specialist.
3. If a thermal method is included in the design, use the thermal method.
4. If a mechanical method is available, and the device can reach the plug, then use the mechanical method.
5. If it is possible to depressurize the system below the hydrate equilibrium pressure, then use the pressure method.

6. If there are chemical injection access points, then use the chemical method.

7. Nonstandard engineering solution is required.

4.7.3.2 Detailed Discussion of Riser Strategy
4.7.3.2.1 Pressure Method
Options
- Two-sided depressurization
- One-sided depressurization

Comments
- Do not increase pressure in the line. This will most likely make matters worse, making the plug harder and more compact, leaving a more difficult plug to remediate.
- After depressurization, the pressure at the hydrate plug should be less than the hydrate formation pressure at ambient temperature for this to be a viable option. Note that this pressure depends on water salinity.
- Keep a fluid cushion downstream of the plug at all times.
- After the plug is melted, ensure that proper amounts of chemicals are applied to prevent the plug from reforming.

Recommendations
- Decrease pressure at facilities (and/or at the base of the riser) slowly. Monitor pressure continuously.

Possible Risks and Concerns
- If a fluid cushion between the hydrate plug and wellhead is not kept, depressurization could cause hydrate plug to travel at high velocities toward the wellhead, creating safety risks and risk of rupturing the riser, or damage to equipment.

4.7.3.2.2 Chemical Method
Options
- Methanol, ethanol, or glycol
- Injection at the facilities
- Injection at an access point

Comments
- Due to density differences, it is preferred to inject glycol (as opposed to methanol or ethanol) at access ports above the hydrate plug to ensure that the chemical reaches the plug.
- For similar reasons, methanol or ethanol should be injected at access points below the plug.

- Keep a fluid cushion downstream of the plug at all times.
- After the plug is melted, ensure that proper amounts of chemicals are applied to prevent the plug from reforming.

Recommendations
- Determine the correct amount of chemical to inject based on free water and released water upon hydrate melting.

Possible Risks and Concerns
- HSE risks associated with chemicals.

4.7.3.2.3 Mechanical Method
Options
- Coiled tubing

Comments
- Keep a fluid cushion downstream of the plug at all times.
- After the plug is dissociated, ensure that proper amounts of chemicals are applied to prevent the plug from reforming.

Recommendations
- Use coiled tubing to jet warm brine or chemical on the hydrate plug.

Possible Risks and Concerns
- HSE risks associated with chemicals.

4.7.3.2.4 Thermal Method
Options
- Electrical heating (using an electric heating ready system)
- Heated bundle (again, already in place)

Comments
- These methods are considered during project design. If they are not currently implemented, other methods must be considered.
- Ensure that the location and length of the entire hydrate plug are known.
- Risks associated with electrical heating need to be assessed.
- After the plug is dissociated, ensure that proper amounts of chemicals are applied to prevent the plug from reforming.

Recommendations
- For either option, careful consideration of pipe wall temperature is essential for safe hydrate remediation. The outer pipe wall temperature corresponding to dissociation pressure that exceeds maximum pipe allowable working pressure, should not be exceeded. Once the plug starts to melt, indicated by pressure increase, the temperature can be ramped up. The heating medium must span the hydrate plug.

- Uniform heating of the pipe wall is imperative.
- Ensure that released gas has a free path to escape.

Possible Risks and Concerns

- If the pipe wall temperature is initially greater than that which corresponds to dissociation pressures greater than maximum allowable pressure, local pressures due to released gas from hydrate melting may exceed bursting pressure of the pipe.
- If the entire plug is not heated uniformly, released gas may not be able to escape, creating high local pressures, possibly exceeding bursting pressure of the pipe.

4.7.4 Equipment Strategy

Hydrates commonly form in a variety of equipment. Valves, separators, manifolds, pipes, heat exchangers, and other similar types of equipment are included in this category. Hydrate plugs in equipment are fairly common and can usually be quickly removed. If the affected equipment is located in relatively inaccessible areas, remediation could be more challenging and therefore takes longer.

Depressurizing the equipment should be the first step. Note that this may not be possible in subsea equipment. If depressurization is not possible, chemical treatment should be considered. Lastly, removal of the plug via mechanical means should be considered as a last resort. An exception to the above will be where thermal management systems are in place to deal with plugs.

4.7.4.1 Recommended Order of Consideration

1. Consult a specialist.
2. If a thermal method is included in the design, use the thermal method.
3. If it is possible to depressurize the system below the hydrate equilibrium pressure, then use the pressure method.
4. If there are access points for chemical injection, use the chemical method.
5. If a mechanical device is available and the device can reach the plug, then use the mechanical method.
6. Nonstandard engineering solution is required.

4.7.4.2 Detailed Discussion of Remediation Strategy

4.7.4.2.1 Pressure Method

Options

- Two-sided depressurization
- One-sided depressurization

Comments
- Do not increase pressure. This will most likely make matters worse, making the plug harder and more compact, leaving a more difficult plug to remediate.
- After depressurization, the pressure at the hydrate plug should be less than the hydrate formation pressure at ambient temperature for this to be a viable option. Note that this pressure depends on water salinity.
- Keep a fluid cushion downstream of the plug at all times.
- After the plug is dissociated, ensure that proper amounts of chemicals are applied to prevent the plug from reforming.

Recommendations
- Decrease pressure slowly to avoid a possible projectile.

Possible Risks and Concerns
- If a fluid cushion between the hydrate plug and the facilities is not kept, depressurization could cause hydrate plug to travel at high velocities, creating risks of equipment damage and/or safety.

4.7.4.2.2 Chemical Method

Options
- Methanol, ethanol, or glycol
- Injection at an access point

Comments
- Due to density differences, injecting glycol (as opposed to methanol or ethanol) is preferred at access ports above the hydrate plug to ensure that the chemical reaches the plug.
- For similar reasons, methanol or ethanol should be injected at access points below the plug.
- Keep a fluid cushion downstream of the plug at all times.
- After the plug is melted, ensure that proper amounts of chemicals are applied to prevent the plug from reforming.

Recommendations
- Determine the correct amount of chemical to inject based on free water and released water upon hydrate melting.

Possible Risks and Concerns
- HSE risks associated with chemicals.

4.7.4.2.3 Mechanical Method

Options
- Coiled tubing

Comments
- Keep a fluid cushion downstream of the plug at all times.
- After the plug is dissociated, ensure that proper amounts of chemicals are applied to prevent the plug from reforming.

Recommendations
- Use coiled tubing to jet warm brine or chemical on the hydrate plug.

Possible Risks and Concerns
- HSE risks associated with chemicals.
- Damage to equipment by mechanical tool.

4.7.4.2.4 Thermal Method

Options
- Circulating warm fluids over hydrate plug
- Heating hydrate plug

Comments
- Ensure that the location and length of the entire hydrate plug are known.
- After the plug is melted, ensure that proper amounts of chemicals are applied to prevent the plug from reforming.

Recommendations
- For either option, careful consideration of pipe wall temperature is essential for safe hydrate remediation. The outer pipe wall temperature corresponding to dissociation pressure that exceeds maximum pipe allowable working pressure, should not be exceeded. Once the plug starts to melt, indicated by pressure increase, temperature can be ramped up. The heating medium must span across the entire hydrate plug.
- Uniform heating of the hydrate is a must.
- Ensure that released gas has a free path to escape.

Possible Risks and Concerns
- If the pipe wall temperature is initially greater than that which corresponds to dissociation pressures greater than maximum allowable pressure, local pressures due to released gas from hydrate melting may exceed bursting pressure of the pipe.
- If the entire plug is not heated uniformly, released gas may not be able to escape, creating high local pressures, possibly exceeding bursting pressure.

4.8 CASE STUDIES

Case studies are difficult to publish. This is because of the commercial impact, and legal and contractual implications as well as the reputation of the operator. Consequently, detailed reports have been rarely published in the past. More recently, however, more events have been at least reported in presentations without formal papers. This has a positive impact on the industry, helping all understand not only how to avoid repeating the event, but also in helping concerned parties in the industry understand how to efficiently remove hydrate blockages at minimum cost and time.

The case studies below are provided with as much detail as permitted or reported in the literature.

4.8.1 Export Pipeline (BP Pompano)

The main Pompano (VK 989) gas export line became plugged with what was subsequently identified as a hydrate blockage. For the previous 4 days the gas dehydration unit on the Pompano platform had not been operating correctly and it was concluded that for 4 days wet gas was produced through the pipeline. During this period, it was assumed that hydrate solids had formed in the pipeline, which progressively accumulated until the plug blocked the flowline.

The quantity of hydrate was approximated by estimating the amount of water injected into the pipeline and by estimating the amount of hydrate detected by densitometer measurements carried out on the pipeline. The ROV-deployable densitometer was a valuable tool that allowed "measurement" of the hydrate plug.

The mass of hydrate based on water injection was calculated as follows:
- Time estimated for "wet" gas production = 4 days
- Average gas flow during this period = 65 MMSCFD (million standard cubic feet per day)
- Water rate = 0.11 bbl water/MMSCF (40 lb/MMSCF)
- Water injected into pipeline = 4 * 65 * 0.11 = 28.6 bbl of water
- Expressed in cubic meters = 28.6 * 0.159 = 4.5 m^3 of water
- Expressed in tons of water = 4.5 tons of water

Assuming that all the water available is converted to hydrate and that hydrate composition is *approximately* 85 wt% water 15 wt% hydrocarbon gas, the maximum hydrate mass was estimated to be 5.3 metric tons (hydrates are 85mol% H_2O and 15% hydrocarbon, but remember these are rough estimates based on coarse measurements). This, of course, is the worst-case scenario; 100% conversion of water phase to hydrate would not be expected.

Hydrate mass was also estimated using pipeline densitometer measurements. Looking at the densitometer readings, it appeared that a hydrate blockage was located approximately 3300 ft from the Pompano platform. The probable hydrates were strewn over about 900 ft in which there was about 100 ft with specific gravity ranging from 0.68 to 0.83, surrounded by another approximately 150 ft in the 0.5 to 0.6 range. The remainder tapered off to the 0.2 to 0.5 range. From the pipeline density measurement it was estimated that approximately 200 ft^3 of hydrate was present in the pipeline, or 5.7 metric tons. Both estimates of the quantity of hydrate were (remarkably) similar, given the levels of uncertainty that were present. The blockage being located approximately 3300 ft from the Pompano platform was consistent with simulations of warm gas cooling and entering the hydrate region some distance from the platform.

4.8.1.1 Strategy Employed to Dissociate the Plug

The recommended procedure agreed upon with the pipeline operator to dissociate the plug was to slowly and very carefully depressurize the pipeline from both sides of the blockage with a bias away from VK 989. Subsea valves downstream could isolate the main line, and access to the downstream end of the plug was available via a line from an adjacent platform MP 313 (Chevron).

While procedures were being agreed upon, methanol injection at VK 989's riser was initiated. Rates were increased as pressures were lowered. A total of 2500 gallons were injected into the riser base. Pipeline pressure was reduced to 500 psi at VK 989, and 300 psi at MP 313. Pressure was then reduced to 200 psi at MP.

Communication across the plug was observed and the pressure at VK 989 had fallen to 480 psi. This pressure value was in good agreement from hydrate model predictions (480 psi dissociation pressure at 55 °F). The rapid communication on either side of the plug indicated that the plug was quite porous and was not extremely dense.

Pressure at VK 989 continued to fall to 400 psi. MP 313 pressure was reduced to 200 psi and communication was again observed at VK 989.

The process was continued until MP 313 was at 100 psi. BK 989 pressure continued to fall to a final pressure of 125 psi. VK 989 was then depressurized to 100 psi to match MP 313. The depressurization procedure was complete after about 28 hr.

The densitometer was deployed and the location resurveyed. The densitometer showed that the plug had shrunk to a 2-ft length, which was considered the last section requiring dissociation. The pressure at VK 989 was increased to 200 psi to attempt to shift the methanol to the plug site, but the plug moved. Pressure was reduced again to 100 psi and the remainder of the plug dissociated after about 62 hr.

The pipeline was then repressurized and production resumed.

By the end of the process, 2900 gallons had been injected into the pipeline at VK 989. Full inhibition with methanol was then applied for the foreseeable future until operators were confident that the glycol unit was performing to specification.

Total downtime was 14 days consisting of 10 days assessing the problem and reaching agreement with other operators on remediation measures, and 4 days depressurizing the flowline to dissociate the hydrate blockage.

4.8.2 Gas Condensate Pipeline (Chevron)

Chevron (Davalath, 1996) reported that a complete blockage formed in a 6-in, 15-mi pipeline. The pipeline was insulated, sufficient to keep the gas above the hydrate formation temperature under flowing conditions. The condensate content was approximately 20 bbl/MMSCF. Although there was no free water, the gas was saturated with water vapor at the pipeline inlet pressure and temperature. Condensed water contributed to forming the hydrate plug. Ambient temperature was about 3 to 5 °C (37 to 41 °F). The blockage occurred during an extended shut-in period over a 300-ft section underneath a road crossing. Hot taps had been used in the past to locate a blockage in the same location. Hot tapping was considered too risky in this case. Hydrates do not typically form in these 6-in lines if the pipeline is depressurized within the first 24 hr.

To remove the blockage, two methods were used sequentially. The line was first depressurized on both sides of the plug. An arc-welding rig was then used to apply electrical current directly to the 300-ft section of the steel pipe. The line was heated to 20 to 25 °C (68 to 77 °F). This approach was effective in melting the hydrate plug. The remedial operation took 2 days to complete.

4.8.3 Well (Gas Condensate)

Davalath and Barker (1993) described a hydrate problem in 595 ft of water located offshore South America. The well was completed with a 7-in casing and 3.5-in tubing. Production consisted of gas and condensate at several hundred barrels per day with a water cut of about 6%. A 15-hr production test was followed by a 25-hr shut-in period to collect reservoir pressure buildup data.

The well was shut in at the surface exposing the tubing to high-pressure gas and cold water (45°F), which resulted in hydrate formation. Under these conditions, the tubing fluid was approximately 29 °F below the hydrate dissociation temperature.

A bridge inside the tubing string blocked wireline tools. Further pulling resulted in separation. The lubricator was subsequently found to be full of hydrates. Attempts were made to melt hydrates by:

- Pouring glycol into the top of the tubing
- Using heated mud and seawater
- Increasing the pressure to 7000 psi at the surface to break the hydrate plug

These attempts were unsuccessful and the authors noted that the pressure increase appeared to create a more stable hydrate, rather than ejecting it from the tubing.

A coiled-tubing string was then run inside the tubing string and 175 °F glycol was circulated to the hydrate plug at 311 ft. Direct contact with the hot glycol removed the hydrate plug.

More than 13 days were lost because of this incident.

4.8.4 Equipment (BP Gulf of Mexico)

At the 2009 PennWell Subsea Tieback Forum and Exhibition in San Antonio, Texas, BP reported that it had successfully removed plugs from a 16-in jumper connecting the Atlantis gas export line to the Mardi Gras gas transport system (Sauer, 2009).

The Atlantis field is located in the Western Atwater Fold-belt along with the 1998 Mad Dog and 1995 Neptune discoveries. Water depths range from 4200 ft on top of the Sigsbee escarpment to greater than 6800 ft off the escarpment. Atlantis is approximately 50 mi southeast of the Troika field and 90 mi southwest of the Mars field. Oil and gas exports from the field enter the Mardi Gras pipeline "Caesar" oil-gathering system and the "Cleopatra" gas-gathering system for transport to market.

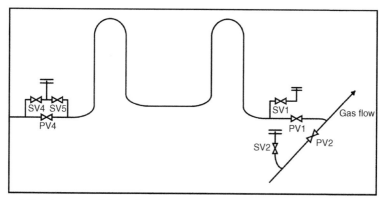

Figure 4.2 Schematic of Atlantis valve arrangement.

Final gas export system commissioning was nearly complete, and full
production was to come on stream soon thereafter. A schematic
(Figure 4.2) of the Cleopatra export line from Atlantis to the system is
shown above. The Mad Dog lateral was in use, transporting gas to shore.

Commissioning was nearly complete for the Atlantis lateral of the
Cleopatra gas export line (Figure 4.3) from the Atlantis production
quarters (PQ) facility to the Mardi Gras pipeline system when the commissioning team discovered a hydrate in the SV-1 service valve on the
W4 pipeline end termination (PLET) just upstream of the Atlantis/Mardi
Gras tie-in point. A gamma ray densitometer survey confirmed hydrate
formation in the J4 jumper (Figure 4.2). Although not stated in the
presentation, the existence of a hydrate deposit implied that line valve

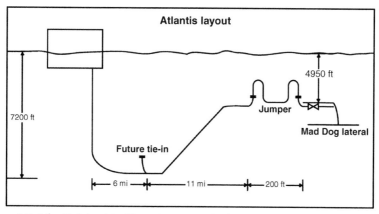

Figure 4.3 Atlantis lateral to Cleopatra gas-gathering system.

PV1 had leaked or was leaking, since the only source of gas to form hydrates would be from there.

Numerous attempts to dissociate the plug using conventional methods, such as vacuum pumps and methanol injection, did not successfully clear the hydrate. Remediation operations were then suspended until a more robust remediation operation could be developed and implemented.

After remediation operations were suspended, a slow but steady pressure increase in the Cleopatra line at the Atlantis PQ facility was detected. When service valve SV-4 at the P4 PLET was closed, the pressure increase at the Atlantis PQ stopped. Subsequent modeling indicated that approximately 9000 barrels of seawater had leaked into the pipeline. The temporary valve stack being used in commissioning operations attached to the SV-4 and SV-5 MEC connector had provided the leak path into the pipeline.

After considering all options, it was decided to use a drill rig to remove fluids from the pipeline, and then chemically treat and depressurize the J4 jumper hydrate.

The only tie-in points were through the 4-in service valves (SV1, SV4, and SV5). To prevent pulling the small piping apart with rig heave, an Atlantis field subsea tree was installed on a temporary conductor near the pipeline jumper. Flexible 2-in pipe was then connected to the production and test hubs from both ends of the jumper via temporary manifolds.

This arrangement then afforded full control and chemical injection capability via the blow-out preventer (BOP) and lower marine riser package (LMRP) stack set on the tree, and fluids management via the workstring, gas lift string, and chemical down line.

The pipeline remediation then consisted of three major components: dewatering the line, chemically drying the line, and jumper hydrate removal.

There were concerns that the void between the two major plugs in the jumper was sealed off and at high pressure (see Figure 4.4). To determine whether there was pressure communication across any of these blockages, specially manufactured strain gauges were installed on either side of the plugs. As pressures changed during the dewatering phase, the strain gauges were monitored by ROV.

Strain gauge measurements indicated that there was pressure communication across the plug, labeled "A" in Figure 4.4, during dewatering. Unfortunately, water was forced through plug "A" resulting in a noncommunicating plug. This was confirmed by the fact that the strain gauges on the other side of

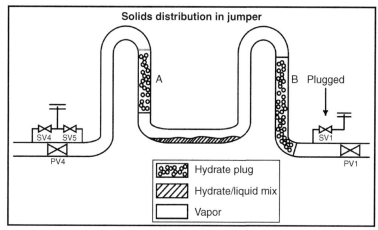

Figure 4.4 Solids distribution in jumper.

"A" stopped responding. The void between plugs "A" and "B" was now sealed at the peak pressure during which water flowed through plug "A."

Following successful dewatering and drying of the lateral, remediation procedures began, whereby pressure was reduced via P4 service piping. At the forum, BP reported that a plug was "gently released" following which the flow path to the workstring was plugged, including the subsea choke.

After the hydrates were cleared from the flow path, two-sided depressurization was initiated to clear the remaining blockages in SV1 and the jumper.

REFERENCES

Bondarev, E.A., Groisman, A.G., Savvin, A.Z., 1996. Adhesive properties of gas hydrates and ice, 2nd International Conference on Gas Hydrates, Toulouse, June 2–6.

Borthne, G., Berge, L., Austvik, T., Gjertsen, L., 1996. Gas flow cooling effect in hydrate plug experiments. Second International Conference on Gas Hydrates, Toulouse, 2–6 June. pp. 381–387.

Davalath, J., 1996. DeepStar IIA CTR A208-1, "Methods to Clear Blocked Flowlines." Mentor Subsea, 157 pages and appendix, January.

Davalath, J., Barker, J., 1993. SPE 26532 Hydrate Inhibitor Design for Deepwater Completions, SPE Ann. Tech. Conf, Houston, TX 3-6 October.

Michel, B., 1978. Ice Mechanics. Les Presses de L'Universite Laval, Quebec. (Adhesive shear strength of ice).

Sauer, S., 2009. Atlantis Hydrate Remediation. Subsea Tieback Forum & Exhibition, March 3–5.

Sloan, E.D., Koh, C.A., 2008. Clathrate Hydrates of Natural Gases, 3rd Ed. Taylor and Francis, Boca Raton, FL.

Streeter, V.L., 1971, Fluid Mechanics, 5th Ed., McGraw-Hill, New York, pp. 140-144.

APPENDIX

Estimate of Potential Damaging Forces in a Bend Due to Sudden Release of Hydrate Plugs

It is necessary to compute maximum forces and maximum pressures imposed by hydrate plugs suddenly released within pipe systems where high differential pressure could be imposed on a plug prior to release/dissociation. Details behind the analytical solution to this problem are given in the following.

System Description

The problem addressed here is that of a single plug bounded by two gas voids. The principle applied here is that only the worst scenarios are assessed to determine the maximum possible damage that could occur during uncontrolled release of a plug at the least opportune moment, which should provide the greatest impact resulting from the risk associated with one-sided or two-sided plug dissociation without fully understanding and controlling the state of the system.

Solution

The system was modeled as a simple mass separating two chambers at different pressures. Figure 4.A1 shows the simple model. Mass M sits at its initial position $x = 0$ at time $t = 0$. The chamber to its right, chamber number 2 is L_2 feet long. Chamber number 1, which is L_1 feet long, binds the mass on the left.

The remaining variables are defined as follows:
P_i = Pressure in chamber i, psia
V_i = Volume of chamber i, ft^3
A = Area of the pipe/jumper, ft^2
m_i = mass of gas in chamber i, lbm
Z_i = Compressibility factor of gas in chamber i

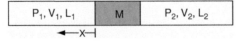

Figure 4.A1 Basic hydrate plug model.

Assumptions

A number of assumptions are made:

- The system is frictionless. This is not completely inappropriate, as the mass will glide on a thin layer of water at the boundary between the mass and the pipe wall (similar to ice skates).
- Change in gas temperature has very little influence on the hydrate mass over the time of interest (sub-second/second response time).
- Average compressibility is constant.
- No leakage of gas around the mass. The mass forms a perfect piston seal against the pipe wall.
- The plug is impermeable to gas.
- The bend is immobile.
- The system possesses sufficient strength to support plug travel. The simulation continues until the plug is halted by high pressure (reverses direction or ruptures the pipe).

$$M\frac{d^2x}{dt^2} + P_1 A - P_2 A = 0 \qquad [4.A1]$$

where

$$P_1(x) = \frac{m_1 Z_1 R T_1}{A(L_1 - x)} \qquad [4.A2a]$$

$$P_2(x) = \frac{m_2 Z_2 R T_2}{A(L_2 + x)} \qquad [4.A2b]$$

Let

$$k_i = \frac{m_i Z_i R T_i}{M} \qquad [4.A2c]$$

Substituting Equations (4.A2) into Equation (4.A1):

$$\frac{d^2x}{dt^2} + \frac{k_1}{(L_1 - x)} - \frac{k_2}{(L_2 + x)} = 0 \qquad [4.A3]$$

Multiply through by $2\frac{dx}{dt}$ and separate:

$$2\frac{d^2x}{dt^2}\frac{dx}{dt} = -\frac{2k_1}{(L_1 - x)}\frac{dx}{dt} + \frac{2k_2}{(L_2 + x)}\frac{dx}{dt} \qquad [4.A4]$$

Integrating

$$\int 2 \frac{d^2x}{dt^2} \frac{dx}{dt} = -\int \frac{2k_1}{(L_1 - x)} \frac{dx}{dt} + \int \frac{2k_2}{(L_2 + x)} \frac{dx}{dt} + C \qquad [4.A5]$$

yields

$$\left(\frac{dx}{dt}\right)^2 = 2k_1 \ln (L_1 - x) + 2k_2 \ln (L_2 + x) + C \qquad [4.A6]$$

@ $t = 0$, $\dfrac{dx}{dt} = 0$, and $x = 0$
Therefore,

$$C = -2k_1 \ln (L_1) - 2k_2 \ln (L_2) \qquad [4.A7]$$

Finally,

$$\left(\frac{dx}{dt}\right) = \sqrt{2k_1 \ln \left(\frac{L_1 - x}{L_1}\right) + 2k_2 \ln \left(\frac{L_2 + x}{L_2}\right)} \qquad [4.A8]$$

The solution to this problem is the gamma function. However, the solution for position versus time is of less interest than velocity versus distance. By substituting values of x, velocity can be evaluated for the full domain of the problem where $-L_2 \leq x \leq L_1$. Further, if one needs position versus time, it can be estimated by directly integrating the velocity versus time locus.

Consider a typical problem for the jumper where there is a 10-ft hydrate plug located between two chambers 50 and 100 ft long with respective pressures of 200 and 1000 psia. When released, the plug will oscillate between the low pressure and high pressure end about the mid-point where pressure would be equal. Since the plug is a perfect piston, and there are no drag losses, the velocity versus distance plot will repeat itself as shown in Figure 4.A2.

Integrating numerically under this curve yields position versus time as shown in Figure 4.A3.

The piston motion of the plug will cause the pressure in the (initially) low–pressure chamber to rise dramatically. The pressure continues to rise well beyond the initial upstream pressure as it begins to absorb the kinetic energy of the plug as shown in Figure 4.A4.

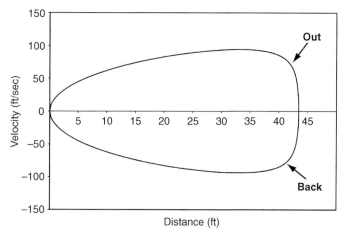

Figure 4.A2 Hydrate velocity plot. Note that top black curved line locus is outbound from initial position, and bottom black curved line is the return path.

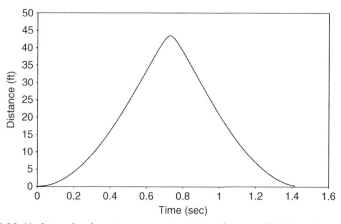

Figure 4.A3 Hydrate plug location versus time. Note that very little time is required to move this rather large mass. In reality, rupture would occur, the plug fragmented, and pipework and structural supports damaged.

Thrust Calculations on a Bend

Thrust is calculated the same way that thrust due to slugging is evaluated in multiphase flow systems. It is simply the resultant vector required to change the direction of a moving mass 90 degrees. The equation used in this analysis follows (Streeter, 1971):

$$F = \sqrt{2}\rho V^2 A \qquad [4.A9]$$

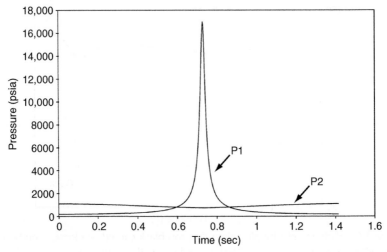

Figure 4.A4 Chamber pressure versus time. Very quickly the pipework receives a high-pressure pulse that is likely to rupture the pipe wall near the end of the low-pressure chamber. In this case it would be a closed valve or another hydrate plug.

where

$F =$ Thrust (lbf)
$\rho =$ Density of the hydrate plug (lb/ft^3)
$V =$ Velocity of the plug as it passes through the bend (ft/sec)
$A =$ Area of the pipe (ft^2)

Velocity and Thrust Evaluation For this example, results show that it is possible to achieve pressures far in excess of 20,000 psia, and velocities ranging from 37 to 270 ft/sec. Thrust due to change in direction of the hydrate mass imposed on a 90-degree bend range from 1200 to 212,000 lbf, depending on the bend location.

CHAPTER FIVE

Artificial and Natural Inhibition of Hydrates

Thierry Palermo and Dendy Sloan

Contents

The objective of this book is to enable the flow assurance engineer to prevent hydrate plugs, and to remediate them once they are formed. The first portion of this chapter introduces inhibition mechanisms of artificial hydrate prevention chemicals as background for the second, larger chapter objective—how natural inhibition of hydrates occurs in some oils, such as those from the Campos Basin, offshore Brazil. The background in this chapter will also enable better understanding of Chapter 6, which is how hydrate kinetic inhibitors can be certified by laboratory tests.

Hydrate inhibition chemicals generally fall into two classes: (1) the traditional thermodynamic inhibitors, such as methanol and monoethylene glycol, and (2) low-dosage hydrate inhibitors (LDHIs), such as kinetic inhibitors and anti-agglomerants. The thermodynamic inhibitors provide a basis for understanding the LDHIs, which in turn provide a background for the naturally inhibited oils. The natural inhibition chemicals, together with shear in the flowline, provide understanding of modern methods of hydrate inhibition. This chapter describes the three parts of hydrate inhibition: (1) thermodynamic hydrate inhibitors, (2) artificial low-dosage hydrate inhibitors, and (3) natural hydrate inhibitors.

Natural Gas Hydrates in Flow Assurance
ISBN 978-1-85617-945-4
DOI: 10.1016/B978-1-85617-945-4.00005-4

5.1 HOW THERMODYNAMIC HYDRATE INHIBITORS FUNCTION AND HOW THEY ARE USED

In Chapter 2, fluid flowline conditions were imposed on the hydrate pressure-temperature diagram shown in Figure 2.1, repeated below as Figure 5.1.

The objective of thermodynamic hydrate inhibitors is to maintain the pressures and temperatures of the flowline fluid (the black "S"-shaped line in Figure 5.1) outside of the gray hydrate region in Figure 5.1. To be effective for the flowline, the hydrate region should be displaced to the left, so that higher pressures or lower temperatures are required to form hydrates. In Figure 5.1, about 23 wt% methanol is required in the free water in the flowline in order to keep the line outside of the hydrate formation region.

From the discussion in Chapter 1, it is clear that hydrates typically form at the interface of the hydrocarbon and liquid water system. In addition to the water that normally resides in the bottom of the flowline due to its density, the top and sides of the flowline circumference are normally water-wet, providing other interfaces for hydrate formation. All water around the flowline generally dissolves thermodynamic inhibitors such as methanol and monoethylene glycol for thermodynamic inhibition.

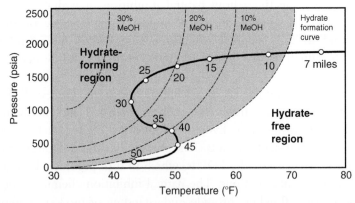

Figure 5.1 Hydrate formation pressures and temperatures (gray region) as a function of methanol concentration in free water for a given gas mixture. Flowline fluid conditions are shown at distances along the bold black curve.

The two common inhibitors, methanol and monoethylene glycol (frequently called ethyleneglycol, glycol, or MEG) have different molecular masses, which cause the injection method of each inhibitor to differ. Methanol, with a low molecular mass (32) is vaporized into the gas phase, where it flows mixed with the gas phase to the point of free water accumulation, usually at a low-lying point on the pipeline, or at the pipe wall as indicated above.

However, because methanol is highly flammable, toxic, and poisons catalysts in downstream systems, MEG is preferred in many parts of the world, such as the Middle East, the North Sea, and the Pacific Rim. The high molecular mass (62) of MEG causes it to be relatively dense and nonvolatile, so there is minimal loss to the gas phase. Instead MEG is usually injected as a liquid to inhibit hydrates in the aqueous phase, as well as to dissolve in water adsorbed at the flowline wall. As a consequence of its high molecular mass, MEG is frequently recovered via removal of water (Sloan, 2000, 26). A comparison of advantages and disadvantages of methanol and monoethylene glycol appears in Table 5.1.

Computer programs such as DBRHydrate®, Multiflash®, PVTSim®, HWHyd, and CSMGem can be used to predict the amount of methanol or MEG required, and how the inhibitor is distributed in each phase—aqueous phase, vapor, and hydrocarbon liquid. While hydrate inhibition normally occurs in the free water phase, a significant amount of methanol is in the vapor and hydrocarbon liquid as well. To calculate the total inhibitor injection rate, one should multiply the inhibitor concentration

Table 5.1 Methanol and Monoethylene Glycol Attributes Comparison

Hydrate inhibitor	Methanol (MeOH)	Monoethylene glycol (MEG)
Advantages	Easily vaporized into gas For flowline and topside plugs No salt problems	Relatively recoverable For plugs in wells and risers Low gas and condensate solubility
Disadvantages	Costly to recover High gas and condensate losses Poisons molecular sieves, catalysts; downstream problems	High viscosity inhibits flow Boiler fouling, salt precipitation

From Sloan, 2000, 26.

in each phase by the flow rate of that phase, and then add the total inhibitor flows in all phases. Methods and examples for the calculation are given by Carroll (2002, 111–123) or by Sloan (2000, 11–19).

Davalath and Barker (1993) reviewed the relative uses of methanol and MEG for typical cases, finding that while the relative injection amounts were approximately the same, the pressure drop was about twice as great for MEG due to its higher viscosity. As another example of the relative use of methanol and MEG, Sloan (2000, 18) presented a case for the relative total amounts of the two chemicals, with respective distribution in the various phases, as given in Table 5.2.

In Table 5.2, it is noted that while significantly less methanol is required in the free water phase for the same inhibition, much more methanol is vaporized into the gas phase, in addition to being dissolved in the condensate, giving approximately the same total amount of inhibitor injected. Methanol is almost never recovered in practice, but is considered an operating cost. Because MEG is recoverable, it is gaining prominence as a worldwide thermodynamic inhibitor.

But how do methanol and monoethylene glycol work to inhibit hydrates? The key to each of the inhibitors is found in their chemical structures, as shown in Figure 5.2.

In Figure 5.2, the inhibition power of methanol and ethylene glycol is due to the attractive nature of the inhibitor oxygen atoms for neighboring water molecules. Each gray oxygen atom in the figure has two lone-pair electrons (Bernal and Fowler, 1933), which provides two negative charges. These negative charges attract the positive charge (on the hydrogen) of a neighboring water molecule to form a strong hydrogen bond between the inhibitor and the water molecule.

Table 5.2 Relative Amounts and Phase Distributions of Methanol and MEG for One Example

	MeOH	MEG
In water, lb_m/MMSCF	174.4	313.1
In gas, lb_m/MMSCF	34.2	0.006
In condensate, lb_m/MMSCF	0.8	0.0061
Total, lb_m/MMSCF	209.4	313.11
Total, gal/MMSCF	31.5	33.3

From Sloan, 2000, 18.

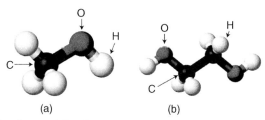

Figure 5.2 Molecular models of (a) methanol and (b) ethylene glycol. The black spheres represent carbon atoms, whites hydrogen, and grays oxygen.

This strong, attractive hydrogen bond between the positive forces of the negative inhibitor oxygen and the positive water hydrogen is also the same force that attracts the oxygen of one water molecule to the hydrogen of another water molecule, in order to form the hydrate cages (Sloan and Koh, 2008, 49–53). The inhibitor hydrogen bond with water may be considered a competitor for the hydrogen bond of water for itself in hydrates, making it difficult to convert all water to hydrates, relative to the case in which inhibitor is not present.

The hydrogen bond, reviewed by Jeffrey (1997), between the inhibitor and the water molecule is very strong, on the order of 10 times the strength of normal van der Waals forces between uncharged molecules. As such, the inhibitor effectively eliminates any attached water molecule from participation in the hydrate structure, since neither methanol nor monoethylene glycol participates in the hydrate cages.

The addition of methanol or monoethylene glycol acts to prevent the water molecules from participating in the solid hydrate structure, but keeps them in the liquid, flowable phase. The more inhibitor added to the system, the more water is prevented from participating in the hydrate structure, so higher pressures and lower temperatures shown in Figure 5.1 are required for hydrate formation from the remaining, uninhibited water.

If thermodynamic hydrate inhibitors such as methanol and ethylene glycol have been effective for the last 75 years, what is the impetus to change? The motivation for change is economics, as exemplified in the recent Ormen Lange presentation by Lorimer (2009).

Production from the Ormen Lange Field, north of the Arctic Circle in Norwegian waters, was started on September 17, 2007. The field produces 70 MMSCMD of gas, accompanied by 430 m^3 of condensed water. The field is 120 km from the Norwegian coast and operates in subzero waters, due to arctic subsea currents. In this flowline 60 wt% MEG is required to inhibit hydrates, and because the flowline temperature is subzero, 10 wt% MEG is required to inhibit ice formation. The amount of MEG injection is very large: 5 to 6 m^3/MMSCF, which requires two 6-in diameter MEG umbilical service lines from the beachhead. The charging of the system with the initial amount of MEG used in this field alone accounts for a substantial fraction of the world's production of monoethylene glycol and has necessitated a significant MEG recovery system.

5.2 THE LOW-DOSAGE HYDRATE INHIBITORS (LDHIs)

Based on the following components, we now consider newer types of chemical inhibitors, called low-dosage hydrate inhibitors (or LDHIs):

1. A qualitative conceptual picture of hydrate inhibition.
2. The understanding that quantitative predictions of inhibition are provided by the aforementioned programs.
3. The motivation for change caused by ever increasing amounts of expensive thermodynamic inhibitors and their recovery.

These modern chemicals fall into two classes: (1) kinetic inhibitors and (2) anti-agglomerants.

Kinetic hydrate inhibitors (KHIs) are low-molecular-weight polymers dissolved in a carrier solvent and injected into the water phase in flowlines. These inhibitors work by bonding to the hydrate surface and preventing significant crystal nucleation and growth for a period longer than the free water residence time in a pipeline. The inhibitor in the water is then removed at a platform or onshore.

Anti-agglomerants (or AAs) are long molecules that cause the water phase to be suspended as small droplets, and rapidly converted to hydrate particles in the oil or condensate. When the suspended water droplets convert to hydrates, flowlines are maintained without blockage, up to about 50% loading in the liquid phase. The emulsion is broken and water is removed in a separator at the end of the flowline.

5.2.1 Kinetic Hydrate Inhibitors

The objective of kinetic hydrate inhibitors (KHIs) is to prevent a hydrate blockage from forming for a time in excess of the residence time of the free water phase in the flowline. KHI performance can be considered as time dependent, unlike the above thermodynamic inhibition chemicals (methanol and MEG), which are time independent. Due to the time

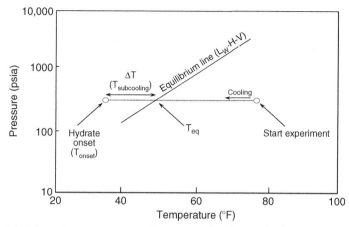

Figure 5.3 Subcooling temperature chart. Note that kinetic hydrate inhibitors (KHIs) are ranked by the degree of subcooling (ΔT) that they can provide below the equilibrium temperature T_{eq} for a given pressure.

dependence of KHIs, another tool is required to determine their effectiveness, the subcooling chart, directly related to the equilibrium time-independent diagram of Figures 5.1 and 2.1.

Figure 5.3 is a pressure temperature plot that is only a portion of Figure 5.1. In Figure 5.3, the diagonal line marked L_W-H-V is a small, straight line of the curved equilibrium line in Figure 5.1 that separates the white, nonhydrate region from the gray hydrate region. The equilibrium line represents the pressure and temperature boundary at which hydrates are stable. As the system cools at a relatively constant pressure, it must not cool beyond the hydrate onset temperature marked T_{onset} in Figure 5.3. At lower temperatures and higher pressures, without a KHI, hydrates will form. Otherwise a hydrate blockage will form due to nucleation and growth of hydrate crystals.

From the recent informal surveys at the SPE Advanced Technology Workshop (Doha, Qatar, January 18–21, 2009), kinetic inhibitors were outselling anti-agglomerants by almost a factor of 2. But how do KHIs work? Again we must turn to the chemical structures, in this case of the four inhibitors (KHIs) shown in Figure 5.4.

Figure 5.4 indicates that the KHIs are polymers composed of polyethylene strands, from which are suspended lactam (with a N atom and a C=O group) chemical rings that are both approximately spherical in shape and polar. The key to the function of these KHI polymers is that they

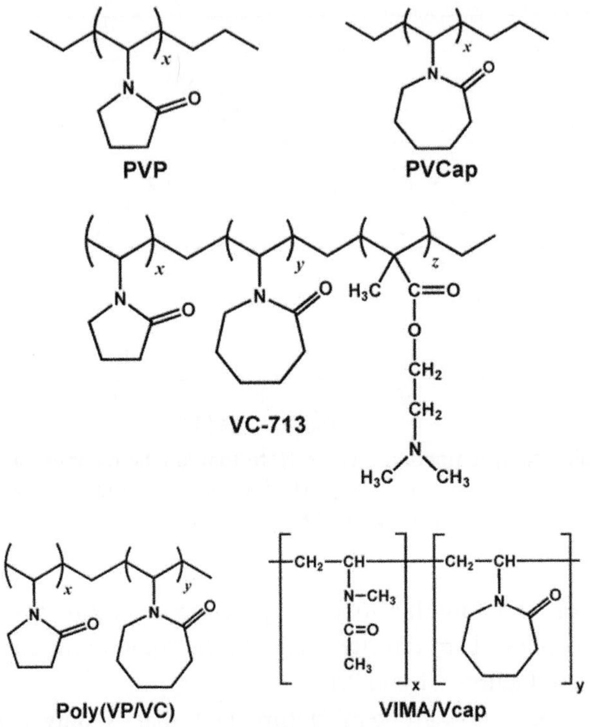

Figure 5.4 Repeating chemical formulas for four kinetic hydrate inhibitors. Every line angle in the figure represents a CH_2 group. The upper horizontal angular line with a repeated parenthesis "$(\ )_{x\ or\ y}$" in each structure suggests that the monomer structure is repeated "x or y" times to obtain a polymer.

adsorb onto the surface of the hydrate, with the polymer pendant group as a "pseudo-guest," in a hydrate cage growing at the crystal surface.

The pendant lactam groups act to "anchor" the polyethylene polymer backbone to the $5^{12}6^4$ hydrate cages surface, and will not allow the polymer to dislodge. So two of the key KHI properties are (1) the pendant group on the polymer must fit into an incomplete, growing $5^{12}6^4$ cage; and (2) the spacing of the pendant groups on the polymer backbone must match the spacing of the growing $5^{12}6^4$ cages on the hydrate crystal surface.

Another way of understanding the way that kinetic inhibitors work is by considering Figure 5.5, in which the open stars can be considered the open $5^{12}6^4$ cages at the hydrate sII crystal surface, shown here to be growing vertically. The closed stars represent the lactam pendant groups attached to the polyethylene backbone of the polymers shown in

$$\Delta T \le \frac{4\sigma}{CL}$$

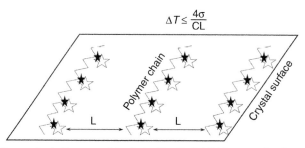

Figure 5.5 Conceptual diagram of hydrate kinetic inhibition mechanism.

Figure 5.4. The lactam groups (filled stars) fit in the $5^{12}6^4$ cages (open stars), anchoring the polymer to the surface of the growing crystal and forcing the hydrate surface to grow past the polymer backbone barrier anchored to its surface (Larsen, 1994).

As the polymer KHI chains are adsorbed more closely together on the crystal surface, it becomes more difficult for the hydrate crystal to grow between them, so the subcooling, or ΔT in Figure 5.3 is increased. The equation relating subcooling to polymer performance is $\Delta T = 4\sigma/CL$, where L is the distance between polymer chains, C is a constant, and σ is the surface energy. This equation indicates that when the polymer chains are closer, the subcooling is greater, because it is more difficult for the hydrate crystal to grow between the polymer chains that are closer together.

5.2.2 Anti-Agglomerants

To understand the mechanism for anti-agglomerants (or AAs), it is important to recall that aggregation is the key for blockage formation in the mechanism from Figure 2.3, repeated here as Figure 5.6.

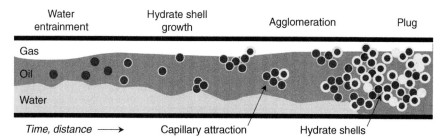

Figure 5.6 Conceptual picture of hydrate formation in an oil-dominated system.

In the Chapter 2 discussion, it was shown that hydrate blockage is the result of four steps shown in Figure 5.6:

1. Water droplets are entrained in the oil phase.
2. At the interface, the droplets grow a hydrate shell to form a "hydrate balloon" of water.
3. The hydrated droplets develop a capillary bridge to aggregate them.
4. With sufficient aggregates, a blockage or plug forms.

If it were possible to prevent hydrates from aggregating, it would be possible for the hydrated droplets to continue to flow with the oil phase, below a reasonable liquid loading such as 60 vol% loading.

As an illustration of the anti-agglomeration concept, at the left in Figure 5.7 is a hydrated droplet that has had an anti-agglomerant, sorbitan monolaurate (Span-20), adsorbed on the surface. The Span-20 caused small, hair-like, fingers of hydrate to protrude into the oil phase so that the two particle attractive forces are smaller than with pure spherical particles, as shown at the right in Figure 5.7. The hydrate particles are kept suspended in the oil phase. It should be noted that the "hairs" at the surface of Figure 5.7 are much larger than the molecules of Span-20 itself.

Typical anti-agglomerants are quaternary ammonium salts, which have one end of the molecule attached to the hydrate structure, and the other end of the long chain dissolved in the oil phase (Kelland, 2006; Sloan and Koh, 2008, 662–668). In this way, anti-agglomerants can be considered as bridge compounds that keep apart the normally aggregated hydrate particles in which they are suspended in the oil phase without aggregation.

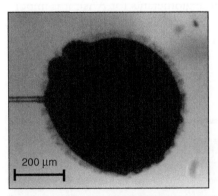 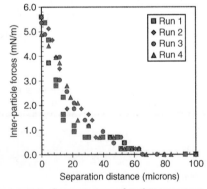

Figure 5.7 Photograph of a hydrate particle grown in the presence of sorbitan mono-laurate (Span-20) at left, and measured forces at right between two such particles as a function of distance. *(From Taylor, 2006.)*

Two more important outgrowths of anti-agglomerants can be indicated. First, as suggested in Chapter 2, if it is possible to prevent a new water phase forming a capillary bridge between the hydrate particles, then the particles will not attract each other, and will flow as in the "cold flow" concept.

Second, for the remaining portion of this chapter, it may be argued that natural hydrate inhibitors work in a similar way to the artificial anti-agglomerants shown here. However, in addition to chemicals it may be possible for both chemicals and shear to prevent agglomeration of the hydrate particles.

5.3 NATURALLY INHIBITED OILS

Polar compounds contained in crude oils, such as asphaltenes, resins, or naphthenic acids, are known to enable stable water in oil (W/O) emulsions to be formed (Førdedal et al., 1996). As previously described, under hydrate formation conditions, crystallization takes place at the water–oil interface, resulting in the building of a solid shell to form a "hydrate balloon" of water. Even for nonsalted water, droplet crystallization is generally incomplete, limited by the gas transfer–limited process through the shell. Whatever the conversion, "hydrate balloons" behave like solid particles and the W/O emulsion progressively evolves to a hydrate slurry. Lab experiments (Sjöblom et al., 2010), loop tests, and field experiences (Palermo et al., 2004) provided evidence for the capability of crude oils to prevent plug formation and to enable transportation. The key issue is to enable the flow assurance engineer to at least anticipate plugging risks, and at best to predict flow conditions under hydrate formation conditions.

In the following, we will present a method to evaluate transportation/plugging conditions of such hydrate slurries. For simplicity, we will limit our discussion to laminar liquid-flow regimes. Also note that this method will only be valid for the water phase totally dispersed in oil as an emulsion. In such a case, it is expected that transportation/plugging conditions are related to the evolution of bulk properties in terms of viscosity of the slurry. In presence of free water in the line, deposit formation might occur. In this case, deployment of prevention methods such as AA injection is recommended. Consequently, in addition to emulsion stability studies, investigation of emulsion formation may be of interest. Indeed, it has been shown that the fraction of water incorporated in the W/O emulsion depends on the initial water cut and the dissipation energy rate. More precisely, the higher the initial water cut, the higher the dissipation energy rate required to incorporate 100% of the water phase into the emulsion (Malot, Noik, and Dalmazzone, 2003).

5.3.1 Viscosity of Suspension

A hydrate slurry can be approximated as a suspension of non-Brownian spherical particles. Viscosity laws of concentrated suspensions are generally expressed as relationships between the relative viscosity μ_r and the particle volume fractions, Φ and Φ_{max}. The relative viscosity is the ratio between the apparent viscosity μ of the suspension and the viscosity of the dispersing liquid μ_0. Φ is the particle volume fraction and Φ_{max} is physically interpreted as the maximum volume fraction to which particles can pack. We will use the equation proposed by Mills (1985), well adapted to hard spheres of equal size and accounted only for hydrodynamic interactions:

$$\mu_r = \frac{1 - \Phi}{\left(1 - \frac{\Phi}{\Phi_{max}}\right)^2}; \quad \Phi < \Phi_{max} = \frac{4}{7} \qquad [5.1]$$

According to Equation 5.1, μ tends to infinity while Φ tends to Φ_{max}. Consequently, there is a theoretical limit in the capability to transport hydrate particles with respect to the maximum volume fraction of about 60 vol% (can be slightly higher for a dispersed phase exhibiting a wide size distribution). However, above a lower value, the suspension does not behave as a liquid anymore; instead it behaves as a paste or a visco-plastic solid, corresponding to a high-risk situation with respect to plug formation. It is difficult to precisely estimate this critical threshold. In a first approximation, the threshold can be set around 50%. By considering the hydrate volume fraction Φ close to the water cut, one obtains the theoretical maximum water cut, which is acceptable in the flowline even if AAs are injected.

5.3.2 Viscosity of Aggregated Suspension

The formation of a hydrate slurry from a W/O emulsion generally results in a rise of the pressure gradient along the line, which in the worst case, can lead to a complete blockage. This phenomenon is due to an agglomeration process during hydrate formation. A hydrate particle is thus an aggregate composed of several "hydrate balloons" formed from individual droplets.

Particles that result from an agglomeration process are generally large and porous. Because of the porosity, the hydrodynamic volume of an aggregate is larger than the total actual volume of the primary particles of the aggregate. Evolution of bulk properties associated with an agglomeration process can be therefore formalized in terms of the evolution of the "effective volume fraction" of hydrate particles Φ_{eff}. The viscosity law of the hydrate slurry becomes:

$$\mu_r = \frac{1 - \Phi_{eff}}{\left(1 - \frac{\Phi_{eff}}{\Phi_{max}}\right)^2} ; \quad \Phi_{eff} < \Phi_{max} = \frac{4}{7} \qquad [5.2]$$

Φ_{eff} can be expressed as a function of the water cut Φ and the average radius ratio between hydrate particles (R) and water droplets (a) according to the following equation:

$$\phi_{eff} = \phi \left(\frac{R}{a}\right)^{3-D} ; \quad R \geq a \qquad [5.3]$$

D is an exponent accounting for the effect of the porosity on the hydrodynamic volume and is commonly named the "fractal dimension." It is a number lower than 3. As an indication, for $D \approx 2$ we can see that Φ_{eff} increases linearly with the average size of hydrate aggregates. For a water cut of 30 vol%, an average radius of hydrate aggregates only twice the radius of droplets, leading to an effective volume fraction of 60 vol% that corresponds to a plugging scenario.

Limitation in the growth of hydrate particles under shear conditions have been experienced (Sjöblom et al., 2010; Darbouret et al., 2008). In the case of a shear-limited agglomeration process, the size of hydrate particles is related to the shear stress τ as follows:

$$\frac{R}{a} = \left(\frac{\tau_0}{\tau}\right)^m ; \quad \tau \leq \tau_0 \qquad [5.4]$$

where τ_0 is the critical shear stress below which aggregates can form and m is an exponent that depends on the breakage mechanism. The critical shear stress is related to the force of adhesion F between primary particles as $\tau_0 \propto \frac{F}{a^2} = \frac{\sigma}{a}$, where $\sigma = \frac{F}{a}$ is the energy of adhesion per unit of area and depends only on the physico-chemical properties of the system.

By combining Equations 5.3 and 5.4, one can write:

$$\frac{\phi_{eff}}{\phi} = \tau_0^X (\tau)^{-X} \quad \text{with} \quad X = (3 - D)m \qquad [5.5]$$

The oil can therefore be well characterized with respect to hydrate agglomeration by the determination of the two physical parameters τ_0 and X. These two parameters can be determined by rheological measurements of the hydrate slurry, either in a high pressure rheometer or a high pressure flowloop, for different water cuts and different shear rates. Providing the viscosity and the shear stress for a large range of condition, τ_0 and X can be calculated according to Equations 5.2 and 5.5. The pressure gradient is a function of the flow rate, and thus the minimum pressure loss to ensure flow can then be easily obtained.

5.3.3 Methodology

At the end of a test during which the hydrate slurry has been formed at a constant rotational speed in a rheometer or at a constant flow velocity in a loop (constant shear rate G), the shear stress τ is recorded (or deduced from pressure gradient measurement in a loop). The end of the test corresponds to the absence of gas consumption and to a shear stress plateau value. By proceeding as follows, experimental results can be represented in the form of Equation 5.5:

- Record the shear stress τ.
- Determine the slurry viscosity $\mu = \tau/G$, where G is the shear rate.
- Determine the relative viscosity $\mu_r = \mu/\mu_0$, where the viscosity μ_0 of the oil phase must be known for same conditions of pressure and temperature.
- Determine the particle effective volume fraction according to Equation 5.2.
- Make a graphical representation of Φ_{eff}/Φ as a function of τ.
- Finally, the two physical parameters τ_0 and X can be deduced by matching Equation 5.5 with experimental results.

In the framework of a project supported by the DeepStar Energy Consortium (CTR 8202), hydrate slurries were investigated in a high-pressure rheometer (Sjöblom et al., 2010). Figure 5.8 shows results for three systems: native Dalia crude oil (Dalia), Dalia crude oil after removing asphaltenes by precipitation with n-pentane (Dalia DAO), and Dalia crude oil after removing the acidic fraction at pH14 (Dalia pH14 extract). The rheometer was a Physica MCR301 (Anton Paar) controlled-stress model equipped with a helix-type mobile. Such geometry was used to avoid settling of the particles and to promote fast dissolution of the gas in the oil.

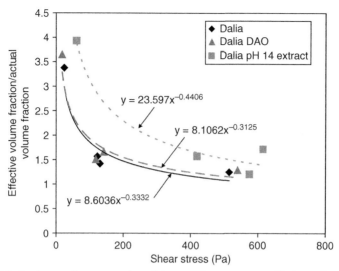

Figure 5.8 Agglomeration properties of three different systems. Hydrate slurries were formed for Φ, G = (0.1, 250 s^{-1}), (0.3, 250 s^{-1}), (0.3, 500 s^{-1}), and (0.4, 500 s^{-1}).

Table 5.3 Value of Parameters τ_0 and X for Three Crude Oils

Crude oil	τ_0 (Pa)	X
Dalia	640	0.33
Dalia DAO	850	0.31
Dalia pH14 extract	1300	0.44

Experimental results can be acceptably represented by power law functions using Equation 5.5. The physical parameters τ_0 and X obtained for each system are reported in Table 5.3.

5.3.4 Prediction

Once the oil is characterized with respect to hydrate agglomeration, flow conditions with the hydrate slurry in a line of diameter d can be predicted by relating the pressure gradient associated to the friction losses $\Delta P/L$ and the velocity u to the shear stress τ and the shear rate G, respectively:

$$\frac{\Delta P}{L} = \frac{4\tau}{d}; \quad G = \frac{8u}{d} \tag{5.6}$$

Finally, for a given pressure gradient $\Delta P/L$, a single value of the velocity u can be determined by successively solving:

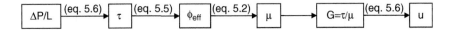

For practical reasons, it may be interesting to evaluate the minimum pressure gradient $\Delta P_{min}/L$ needed to ensure the flow when the hydrate slurry is forming. If we assume that a risky situation is achieved if Φ_{eff} becomes higher than 50 vol%, the minimum shear stress τ_{min} is

$$\tau_{min} = \tau_0 \left(\frac{\phi}{0.5} \right)^{1/X} \tag{5.7}$$

Therefore, the minimum pressure gradient is

$$\frac{\Delta P_{min}}{L} = \frac{4\tau_0}{d} (2\phi)^{1/X} \tag{5.8}$$

Evolution of the minimum pressure gradient with the water cut for different line diameters is illustrated in Figure 5.9 for a crude oil such as Dalia. Note the strong impact of the water cut.

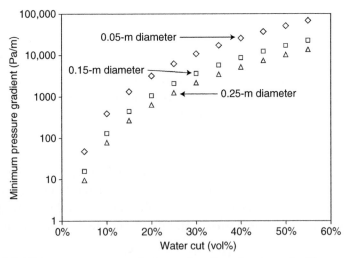

Figure 5.9 Minimum pressure gradient as a function of water cut with $\tau_0 = 640$ Pa and $X = 0.33$.

5.4 CONCLUSION

This chapter first indicated the concept for normal thermodynamic inhibition by methanol and MEG—namely by competing for available water. Then two new low-dosage hydrates inhibitors were discussed: kinetic hydrate inhibitors (KHIs) and anti-agglomerants (AAs). It was shown that KHIs inhibit hydrates by preventing crystal nucleation and/or growth via adsorption at the growing surface of the crystal. In contrast, AAs rapidly convert all emulsified water to hydrates, but one end of the AA attaches to the hydrate crystal, while the other has affinity for oil, keeping the small hydrate crystal suspended in the oil without agglomeration.

The final and most important portion of the chapter provides a quantitative description of how naturally inhibited oils function, showing that the key to anti-agglomeration is provided both by surface chemistry and by shear. Using the methods of this chapter, the engineer can predict the minimum pressure drop required to provide flow in a line that includes suspended hydrates. The evidence for these quantifications was provided both by laboratory apparatuses and flowloops.

In Chapter 6, we will propose a standard method to show how laboratory apparatuses can be used to quantify the performance of KHIs, with translation first to flowloops and then to flowlines.

REFERENCES

Bernal, J.D., Fowler, R.H., 1933. *J. Chem. Phys.* 1, 515.

Carroll, J., 2002. *Natural Gas Hydrates: A Guide for Engineers.* 270 pages, Gulf Publishing, Tulsa, Oklahoma.

Darbouret, M., Le Ba, H., Cameirao, A., Herri, J.M., Peytavy, J.L., Glénat, P., 2008. Lab Scale and Pilot Scale Comparison of Crystallization of Hydrate Slurries from a Water-in-Oil Emulsion Using Chord Length Measurements, Proceedings of the 6th International Conference on Gas Hydrates, Vancouver, Canada.

Davalath, J., Barker, J.W., 1993. Hydrate Inhibitor Design for Deepwater Completions, Proc. 68th Annual Tech. Conf. of Society Petroleum Engineering, SPE 26532, Houston, Texas, October. pp. 3–6.

Førdedal, H., Midttun, Ø., Sjöblom, J., Kvalheim, O.M., Schildberg, Y., Volle, J.L., 1996. A multivariate screening analysis of W/O emulsions in high external electric fields as studied by means of dielectric time domain spectroscopy, II: Model emulsions stabilized by interfacially active fractions from crude oils. *J. Colloid Interface Sci.* 182, 117–125.

Jeffrey, G.A., 1997. *An Introduction to Hydrogen Bonding.* Oxford University Press.

Kelland, M.A., 2006. History of the Development of Low Dosage Hydrate Inhibitors. *Energy and Fuels* 20 (3), 825.

Lorimer, S., 2009. MEG for Hydrate and Ice Control: Ormen Lange Experience SPE Advanced Technology Workshop, January 18–20, Doha, Qatar.

Malot, H., Noik, C., Dalmazzone, C., 2003. Experimental investigation on water-in-crude oil emulsion formation through a model choke-valve: droplet break-up and phase inversion. 11th Int. Conf. on Multiphase 03, June 11–13. Published by BHR Group Limited, San Remo, Italy.

Mills, P., 1985. Non-Newtonian behaviour of flocculated suspensions. *Journal de Physique Lettres* 46, L301–L309.

Palermo, T., Mussumeci, A., Leporcher, E., 2004. Could hydrate plugging be avoided because of surfactant properties of the crude and appropriate flow conditions, OTC. May 3–6. Houston, Texas, Paper n°16681.

Sjöblom, J., et al., 2010. Investigation of the hydrate plugging and non-plugging properties of oils. *Journal of Dispersion Science and Technology* Vol 31, pp 1–20.

Sloan, E.D., 2000. Hydrate Engineering. In: SPE Monograph, vol. 21. Society of Petroleum Engineers, Inc., Richardson, Texas.

Sloan, E.D., Koh, C.A., 2008. *Clathrate Hydrates of Natural Gases*, third ed. Taylor and Francis, Inc., Boca Raton, Florida.

Taylor, C.J., 2006. Adhesion Force Between Hydrate Particles and Macroscopic Investigation of Hydrate Film Growth at the Hydrocarbon Water Interface, Master of Science Thesis. Colorado School of Mines, Golden, CO.

CHAPTER SIX

Kinetic Hydrate Inhibitors Performance

Mike Eaton, Jason Lachance, and Larry Talley

Contents

6.1 INTRODUCTION

Many studies have been conducted to predict kinetic hydrate inhibitor (KHI) performance in gas-dominated field applications. The main questions addressed in this chapter are what test apparatuses and what test protocols adequately predict field performance.

Natural Gas Hydrates in Flow Assurance
ISBN 978-1-85617-945-4
DOI: 10.1016/B978-1-85617-945-4.00006-6

105

The conclusion from this study is that KHI performance in turbulent flow pipelines can be successfully predicted in closed system test apparatuses between 0.5-in diameter to 4-in diameter. Hydrate appearance times measured in turbulent flow/turbulent mixed test apparatuses correspond to hydrate appearance times measured in turbulent, flowing pipelines.

The method of determining hydrate appearance times in the laboratory are described in this chapter. The time from when a rapidly cooled fluid is at a constant temperature below the hydrate equilibrium temperature and when hydrates first appear is called the hold time. This work prescribes that hold times measured in lab apparatuses be made under turbulent mixing conditions only and that the results be used to predict hold times in turbulent flowing pipelines only. Hold times measured during shut-in or in stratified flow pipelines and during shut-in or in stratified flow test lab apparatuses, including rocking cells (Klomp, 2008), have been observed to deviate significantly from hold times measured in turbulent, flowing systems. These observations should serve to caution the reader to consider flow regime as a very significant factor in measuring and predicting KHI performance in test apparatuses and pipelines.

Throughout the hydrate community, different testing apparatuses and procedures are used to test KHI effectiveness. Typically, ExxonMobil performs KHI tests in high-pressure miniloops (HPMLs). It has been found that testing conditions and procedures can have a significant effect on KHI performance. Over several years of testing, procedures have been implemented that give accurate, reproducible results for KHI hold times in the miniloops. These procedures will be illustrated through the results of a laboratory study conducted in miniloops by ExxonMobil Upstream Research Company on KHIs from different chemical suppliers, as well as KHIs from in-house synthesis.

6.2 STUDY 1: MINILOOP FLOWING KHI HOLD TIME

This study (Talley, 2003) observed simple chemical kinetics for the first appearance of hydrates in a natural gas system containing kinetic hydrate inhibitors (KHIs).

KHIs were tested in a flowing miniloop with fluids comprised of a synthetic gas mixture, a stabilized natural gas condensate and fresh water (lab tap water). The hold time, or time to first indication of hydrate formation, was determined as a function of KHI concentration and subcooling. The results fit an exponential function of the natural log of the reciprocal of the hold time versus subcooling.

The experimental work was done at a single constant pressure, 2250 psig (15.5 MPa). The observed turbulent flow hold times ranged from 10 min to 900 hr. The KHI concentrations ranged from 0.025% to 0.30% weight of active polymer, based on the water phase. The sub-coolings ranged from 9 to 25°F (5 to 13.9°C).

The procedure for measuring hold time was to first determine the equilibrium dissociation temperature of the fluids in the loop before adding any KHI. The dissociation temperature was determined at least twice for every charge of fluids.

Figure 6.1 shows the (1) rapid cooldown of bath and fluid temperatures, hydrate formation at 1.3 hr determined by (2) increased differential pressure across the loop circulation pump, (3) temperature increase in the gas transfer line, (4) fast programmed warmup followed by (5) slow warmup, (6) fast drop in differential pressure across the circulation pump at 4.49 hr, (7) fast decline of gas transfer line temperature, and (8) final warmup at the end of the cycle. Hydrate equilibrium dissociation occurred at 4.49 hr at a bath water temperature of 69.3°F and average loop fluid temperature of 70.1°F.

Next, a KHI was injected at the lowest concentration and onset experiments were performed. An onset run is where the miniloop is cooled monotonically until hydrate formation is detected. The loop was cooled from above the hydrate temperature at a rate of 6°F per hour. The bath temperature at which hydrate formation occurred rapidly was determined at least twice.

Figure 6.2 is an example of an onset run. In this case, hydrates were first detected at 5.8 hr by the observation of noise in the differential pressure curve (pump ΔP). The differential pressure is measured across the gear

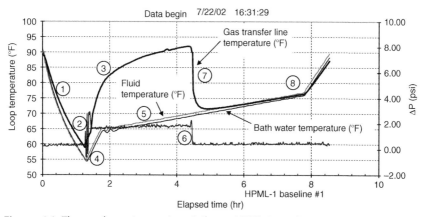

Figure 6.1 Thermodynamic run in miniloop. HPML-1 studies, temperature history: 500 ml seawater, 400 ml condensate at 1000 psig gas.

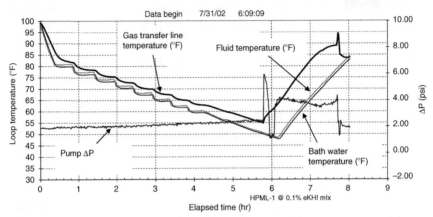

Figure 6.2 Hydrate onset run in miniloop. HPML-1 studies, temperature history: 500 ml seawater, 400 ml condensate at 2250 psig (at 90°F) gas.

pump that circulates the fluids in the miniloop. Slight changes in the apparent viscosity of the fluids result in changes of the differential pressure. Hydrates cause an erratic change in the viscosity and the differential pressure before there is any visible indication of hydrate formation. The hold time for an onset run such as this is typically set equal to 0.1 hr, a default value for a short hold time that is not precisely determined.

After measuring onset temperatures, the hold time was measured for various values of subcooling. The miniloop was cooled from above the hydrate temperature to a specific temperature between the onset temperature and the equilibrium temperature. The bath temperature was held isothermally until hydrate formation was detected. The amount of time that the contents of the loop were hydrate free at the isothermal temperature was defined as the hold time.

In normal determinations of the hold time, the miniloop temperature is lowered to a fixed temperature at the same cooldown rate as the onset run, but hydrate formation is slow and does not occur for hours or days. Figure 6.3 is an example of a 125-hr hold time observation. The first sure indication of hydrate formation is the erratic behavior of the differential pressure starting at 130 hr.

Some of the hold-time runs resulted in hydrate formation rates that were so slow that it was necessary to terminate the run before hydrate formation was observed. When this occurred, the data point was included in the figures as a data point at the last observed hold time with an arrow pointing to longer hold times to indicate that hydrates had not formed.

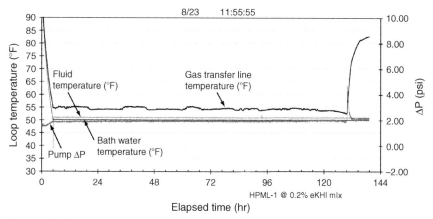

Figure 6.3 Flowing hold-time run in miniloop. HPML-1 studies, temperature history: 500 ml seawater, 400 ml condensate at 2250 psig (at 90°F) gas.

Figure 6.4 is a semi-log plot of (1/hold time) versus subcooling for three concentrations of KHI polymer. These results can be summarized by three linear functions that have a common slope at all KHI concentrations and subcoolings tested.

It was assumed that the true functions of (1/hold time) versus subcooling for different KHI concentrations cannot cross each other. Therefore, in this study it was assumed that the true function is best fit by linear functions that are parallel. The best fit to the data was obtained

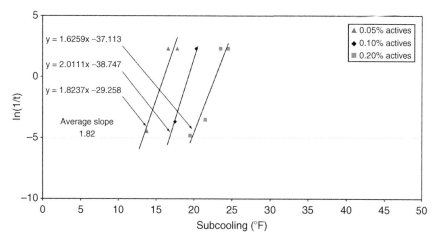

Figure 6.4 Correlation of hold time and subcooling: ln (1/hold time) versus subcooling, eKHI in HPML-1, at 2250 psig, 500 cc fresh water/400 cc gas condensate.

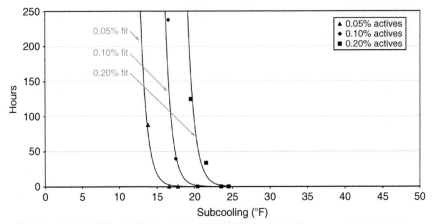

Figure 6.5 eKHI hold time versus subcooling: eKHI in HPML-1, at 2250 psig, 500 cc fresh water/400 cc gas condensate.

by regressing each set of data for each KHI concentration to determine the slope of the line at each concentration. The slopes were averaged to obtain one common slope. The data for each concentration were then fit to the regression line by adjusting only the value of the intercept. This technique is called the forced intercept method.

The resulting curves for the functions derived in the procedure described above were calculated in a spreadsheet and plotted over the experimental hold time data for three KHIs. The calculated results and the corresponding experimental data are shown in Figures 6.5 to 6.7.

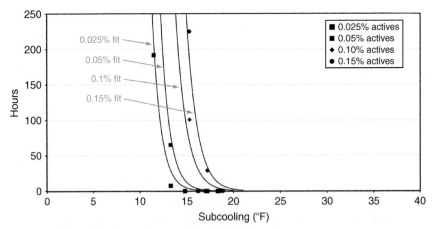

Figure 6.6 KHI-1 hold time versus subcooling: KHI-1 in HPML-2, at 2250 psig, 500 cc fresh water/400 cc gas condensate.

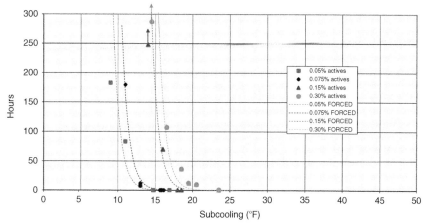

Figure 6.7 KHI-2 hold time versus subcooling: KHI-2 in HPML-2, at 2250 psig, 500 cc fresh water/400 cc gas condensate.

The plots of hold time versus subcooling show the same exponential correlation to subcooling for all KHIs. The hold time decreased exponentially as the subcooling of the isothermal temperature step increased. This behavior has been observed in other gas condensate systems, including nonacid gas systems.

The simple exponential relation observed for these experiments are affected by certain species (e.g. surface active species, such as surfactants, corrosion inhibitors, drilling fluid additives, etc.) not present in these experiments. The kinetics become complex and do not fit the simple exponential relationship when these species are present. Some of these effects are presented in Study 2.

6.3 STUDY 2: AUTOCLAVE TESTING METHODOLOGY

While flowloop testing is the primary method used by ExxonMobil for kinetic hydrate inhibitor (KHI) qualification, quality assurance (QA), and quality control (QC), flowloops are both expensive and complex pieces of equipment. There is a need for a simpler, yet experimentally comparable testing apparatus. Such a piece of equipment, a high–pressure autoclave, was found to repeat miniloop results in terms of both accuracy and precision on KHI hold times using multiple testing conditions, multiple fluid types, and multiple KHIs.

This section presents the basic design and operation of such an autoclave, and the means by which KHI performance comparable to that seen in a miniloop can be ascertained in the device. This methodology will allow KHI selection and QA/QC work to be performed quicker and at less cost than with flowloops.

It was found that a 300-ml vessel, filled with 100 ml of water, 100 ml of gas condensate, and 100 ml of vapor, stirred at a rate of no less than 600 rpm, and controlled to within ~0.2°C of the desired testing temperature successfully reproduced miniloop KHI behavior to experimental precision.

Although efforts were made to demonstrate that KHI QA/QC work in an autoclave can be performed to the standard of a flowloop, the variety of oilfield chemicals and fluid compositions precludes the statement that "autoclaves always yield the same results as flowloops do." Autoclaves cannot effectively simulate the variety of fluid flow regimes that flowloops do, and therefore will not be able to reproduce flow regime-dependent chemical behavior. However, the inhibitory performance of KHIs will be shown to match miniloop results for one application.

6.3.1 Introduction

An autoclave is an apparatus in which special conditions (high or low pressure or temperature) can be established for a variety of applications. Autoclaves make an ideal vessel for fast hydrate inhibitor hold time determination. Figure 6.8 is based upon hydrate equilibrium predictions of the gas A and B

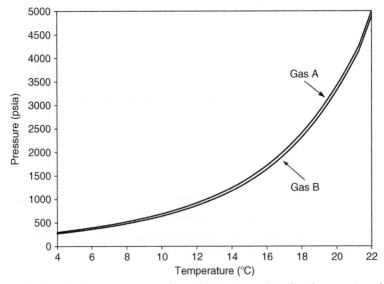

Figure 6.8 Predicted hydrate curves for condensate combined with gases A and B.

Table 6.1 Experimental Vapor Compositions

Component	Gas A mole %	Gas B mole %
H_2S	0.0	0.746
CO_2	3.103	2.357
Nitrogen	3.919	3.919
Methane	85.723	85.723
Ethane	4.765	4.765
Propane	1.648	1.648
n–butane	0.324	0.324
i–butane	0.518	0.518

compositions (synthetic vapor) shown in Table 6.1, combined with a condensate. This report serves to identify a means by which KHI performance may be assessed in an autoclave in comparison to a miniloop. Because H_2S has not been used in an ExxonMobil hydrate miniloop, and because the presence of H_2S imposes both material constraints and additional safety considerations, both gases were used in this study and their functional equivalence will be shown.

Although the two curves shown in Figure 6.8 represent two distinct gases: one with and one without hydrogen sulfide, the hydrate curves lie very close to each other. For a given system pressure (e.g., 2250 psia), the equilibrium temperature, T_{eq}, has a difference of only 0.2°C, which is within the expected error on the prediction and, which will be shown, the experiment. More importantly, the substitution of CO_2 for H_2S in this system has a very minimal impact on the concentration of acid-gas species in the aqueous phase; there is a fluid equilibrium-predicted 0.01% increase in the aqueous acid gas concentration for the H_2S-containing vapor system.

It has been observed that higher acid gas (CO_2, H_2S) concentrations in a system have an antagonistic effect on kinetic hydrate inhibitor performance. That is, for a given subcooling, systems with higher CO_2 and H_2S concentrations have a shorter kinetic hydrate inhibitor hold time than systems that lack or have lower concentrations of such species. Thus it was hypothesized, and will be shown, that a substitution of CO_2 for H_2S in a KHI hold-time test has a negligible effect on KHI behavior. Also to be shown is that the validity of this hypothesis has a significant positive impact on material selection and safety concerns when performing KHI quality assurance and quality control (QA/QC) tests.

To be considered to have the equivalent to the miniloop, the auto-clave, or any other testing apparatus, must satisfy several basic criteria:

1. KHI failure in the system must be able to be experimentally demon-strated as in the miniloop, with a simultaneous temperature increase (due to the exothermic nature of hydrate formation) and pressure or volume drop due to rapid gas consumption. A pressure drop would be seen in a closed, isochoric system. A volume decrease, as in the miniloop, would be seen in an isobaric, constant pressure, system.
2. Hold times, as a function of temperature (subcooling), must not be statistically significantly different than those obtained in a miniloop for the same conditions.
3. Standard deviations on hold times for a given subcooling must not be so large as to render such a comparison inconclusive.
4. Temperature and pressure control must be similar to or better than the miniloop.

With these requirements in mind, the autoclave device and test procedures were designed.

6.3.2 Miniloop Equivalence Requirements

Being able to discern the onset of hydrate formation through a temperature spike and gas volume drop is, perhaps, the easiest of the requirements for functional miniloop equivalence from an instrumentation standpoint, but is most difficult from an experimental standpoint. Figure 6.9 illustrates hydrate

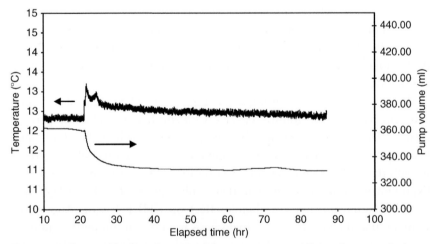

Figure 6.9 Observed hydrate formation via a temperature spike and pressure/volume drop in an autoclave.

formation behavior in a system that is inhibited with a KHI. The system in question, gas "A," condensate, and water, was held at 2250 psia through the use of a pump that was programmed to add or remove gas as necessary to maintain pressure. In this case, pump accumulator *volume* would change in response to gas consumption. The *theoretical* formation temperature for this system, from Figure 6.8, is 17.8°C. As shown, the system was held at approximately 12.3°C for approximately 20 hr with KHI. Without the presence of KHI, the mixture would hydrate extremely quickly (<0.1 hr).

A rapid temperature rise at a time of 20 hr coincided with a precipitous pump accumulator volume drop. (Note: A weak accumulator volume decline prior to the events at hour 20 is ascribed to supersaturation observed by Fleyfel et al. [1993], and others.) These two *simultaneous* events are the hallmarks of hydrate formation in a closed system in which formation cannot be visibly observed. The reason for the subsequent volume stabilization and temperature return to 12.3°C is due to the fact that formation stopped within the system at a time of 30 hr.

A temperature rise in the system may only be observed if the mass of water, and hence the mass of hydrate formed, is sufficient to raise the temperature of the vessel (miniloop, autoclave, or otherwise), an amount such that it can be observed with common laboratory equipment, such as a thermocouple or a resistance temperature device (RTD). This might lead to the conclusion that a system that is mostly water with very little gas or condensate might be most useful. However, using excessive water (or oil) precludes a useful volume of vapor inside the system. Maintaining a chemically homogenous and compositionally stable vapor phase is essential in these tests, as a shifting composition or nonwell-mixed vapor will yield inconsistent hydrate equilibrium temperatures, thus altering perceived driving force. Additionally, having a large volume of liquid, due to the increased solubility of gases at lower temperature, would require a large reservoir of gas to ensure complete saturation of both the condensate and the water under pressure. While no size constraints were initially placed on a "miniloop equivalent test," constructing a large apparatus defeats, to some extent, the purpose of finding miniloop alternatives. Finally, hydrate formation in such apparatuses requires vigorous agitation to ensure renewed water/hydrate former exposure and to observe a "true" (i.e., nonmetastable) hydrate behavior (Sloan and Koh, 2008, 327). In unmixed or nonwell-mixed systems, severe mass transfer limitations are imposed on hydrate creation, and thus formation occurs not as a catastrophic hydration point but as a practically imperceptible rate of gas consumption. As such, rapid water/hydrocarbon contact required that excessive amounts of water not be used. It was decided, based on these criteria and a literature survey that equal volume fractions (at standard conditions) of water and stabilized condensate would be used, with the remaining volume to be occupied by high-pressure gas (Lederhos et al., 1996; Anderson, 2002; Turner, 2005).

Temperature control comparable to that in a miniloop must be matched in order to ensure a proper comparison of the results from each apparatus. *This factor is paramount.* Hydrate hold times at constant KHI concentration have been shown to decrease exponentially with increasing subcooling. Because of this, even small thermal fluctuations, especially at high subcoolings, can cause difficulty in interpretation and give erroneous results. The ExxonMobil miniloop has shown a standard deviation in temperature control of approximately 0.16°C (0.3°F), and is therefore the standard by which a comparable device is held.

Pressure control in an isobaric system (or pressure *monitoring* in an isochoric system) is also important, but the accuracy of such control/monitoring is typically far better than is required when considering general hydrate behavior. The reason for this is as follows. The ExxonMobil miniloop HPML-3 has shown a standard deviation on pressure control of approximately 5 psi. At 2250 psia, this corresponds to a hydrate equilibrium temperature prediction difference of approximately 0.017°C (0.03°F), or 10 times the precision required on the temperature control. With this in mind, pressure control during an isobaric test within 50 psi (for the temperature ranges studied) of the desired setpoint will give hold-time results comparable to miniloop performance.

6.3.3 Device Design

Two contract laboratories and their autoclaves, Lab A and Lab B, were used in the studies comparing autoclave data to flowloop data. The reaction vessels used by each, although slightly different in overall dimension, have specifications close to the standard 300-ml bolted closure reactor from Autoclave Engineers (2009a), a common type of high-pressure laboratory reactor, and one that has been used for over 20 years in hydrate studies (Lederhos et al., 1996; Anderson, 2002; Turner, 2005; Englezos et al., 1987). The units have an internal rounded bottom such that the total internal volume is nominally 300 ml. As will be shown, slight deviations in such geometries and internal volumes were not found to impact the results of hold-time tests significantly. Most important, as will be discussed, was the stirring rate within the system, and the lack of "dead legs" in the tested volume.

While Lab A and Lab B autoclaves are of similar design, some modifications to each design, outlined herein, were made over the course of testing. These changes were minor and nonstructural, and allowed for greater control and increased precision on the measurements. Such changes are outlined in this report and the description of the device, rather than the as-manufactured drawings, should be followed for proper compliance with ExxonMobil-developed autoclave testing methods for KHIs.

Autoclaves purchased from Autoclave Engineers have two sections: the main reaction body and the reactor "head" that bolts or screws onto the body section and contains a stirring rod, impeller, and ports for various measurements and/or injections. These will be discussed in more detail later. A thermocouple is used for monitoring the internal vessel temperature.

Two angled ports in either side of the apparatus were used for pressure monitoring and chemical injection.

Other than the requisite temperature and pressure control mentioned earlier, there are two important design factors that allow the acquisition of accurate and precise hold times in an autoclave:

1. Elimination of the potential for stagnant condensed water or vapor
2. Adequate mixing to ensure elimination of mass-transfer limitations in hydrate formation, and "splashing" to ensure deposition of KHI to all spaces within the system (fluid homogenization)

Autoclaves are often sold with a number of head ports with attached tubes that dip into the liquid volume and allow sampling or gas sparging to be performed. Such accessories are to be avoided in KHI tests, as these have been found throughout the course of ExxonMobil experimentation to cause tests to fail prematurely. The reason for this is that nonwell-mixed water, containing insufficient amounts of KHI and gas can form hydrates within such a tube. The creation and propagation of the hydrate seed then causes the entire system to fail, as KHIs cannot abate the growth of an already-formed hydrate. The only exception to this is the presence of a thermowell. The thermowell is a closed-end tube that is placed inside the system near the bottom of the body and allows temperature measurements to be made without direct immersion of the temperature monitoring device in the process fluids. The standard length of the thermowell, 0.133 m (5.25 in), was found to provide a suitable means by which the temperature in the system could be monitored and controlled.

Elimination of mass transfer limitations in the system necessitates that the amount of gas entering the water or condensate at a given stirring rate would not limit the rate of hydrate growth. In other words, hydrate growth should only be limited by the intrinsic rate of reaction, and not the availability of the reactants. This is important in KHI hold-time tests because in the event that a system is mass transfer limited, hold-time results will be erroneously long. KHI performance would be overpredicted, and thus dosages would be inappropriately low. Mass transfer of the vapor to the liquid is enhanced by high rates of agitation and stirring up to a certain point, at which time kinetic limitations to hydrate formation take over. The lower bound on stirring rates is shown in the work by Herri et al. (1996). At low stirring rates (<250 rpm), the observed rate of gas consumption in the system was negligible, owing to the fact that gas hydrate formation stopped once hydrate formers in the liquid were exhausted, and that low stirring rates did not allow new vapor to be absorbed into solution.

The upper bound on stirring rates (i.e., those that achieve kinetic limitations on hydrate growth) were seen by Englezos et al. (1987). The "measured" kinetic rate parameter was shown to increase, as expected, with increasing stirring rate from 250 to 400 rpm. This indicates that for this range of stirring, the hydrate reaction kinetics observed are not the true, intrinsic kinetics. Instead, as mentioned previously, they are the kinetics observed with coincident mass transfer limitations. However, for stirring rates in this system, which system was similar in configuration and size to the systems tested at Lab A and Lab B, above 400 rpm, the rate constant for the system was truly constant. This indicates that at such rates, the system is no longer mass transfer limited, and is capable of displaying behavior similar to that of the miniloop and, more importantly, the pipeline.

In addition to the elimination of mass transfer limitations, it was also essential to ensure complete interior wetting of the autoclave system to prevent condensed water from hydrating and causing the KHI hold-time test to fail prematurely. It was found experimentally using two-thirds the system volume of pure water (200 ml) that a stirring rate of 600 rpm caused the liquid level to reach all points on the inside of the system. This was verified by inserting a sample tube at the uppermost portion of the autoclave head and attempting to withdraw fluid. The withdrawn sample contained a large amount of liquid, and was therefore "wetted." In a typical test, one half the liquid volume would be comprised of gas condensate, but as this was a proof-of-concept test, only water was used. Note that the required stirring rate is geometry dependent, and thus this rate is only valid for the system described herein.

It is recommended that, at minimum, the stirring in the system be kept above 600 rpm. Stirring rates as high as 1300 rpm were found to have no significant difference from results obtained at 600 rpm. Of course, mixing in the system is geometry dependent, not only in the geometry of the cell itself, but also on the geometry and location of the impeller. Two impeller systems were tested in the verification of this device. One, the Lab A, uses a 0.022-m (0.88 in) diameter shaft connected to a 0.032-m (1.25 in) diameter, six-blade Dispersimax turbine (Autoclave Engineers, 2009b).

The Dispersimax paddle-style impellers are often packaged with a hollow shaft that draws vapor from the vapor space and disperses it throughout the liquid in an effort to increase mixing and mass transfer. These hollow shafts should be avoided, as they create a potential area for improper mixing, leading to early test failure or inconsistent results. A solid shaft is ideal. The standard 300-ml mounting of the impeller at the base of the 0.17-m (6.69-in) shaft was found to be suitable for KHI hold-time testing.

The Lab B configuration used a custom two-blade mixer on a solid shaft. As will be shown, hold-time results using these mixers displayed good agreement with miniloop data. It appears that flat-blade impellers, which give both radial and axial mixing, are suitable to meet the KHI hold time criteria set forth at stirring rates above 600 rpm. Magnetic stir bars, which are common in atmospheric-pressure reaction vessels, should never be used, as the degree of agitation and shear they provide is insufficient to ensure adequate mixing.

The final and most important design criterion that needed to be met was that of temperature control in the system. As mentioned, temperature stability is of utmost importance due to the exponential nature of KHI hold times with subcooling. Autoclave systems are often sold with internal cooling coils that circulate a heat transfer fluid such as a mixture of ethylene glycol and water within the coil. However, prior experience has shown that such a process control scheme can lead to poor temperature control within the system, such as temperature overshoots, thermal oscillations, and premature KHI failure. Therefore, the preferred method of cooling or warming of the system should be done externally using an insulated glycol/water bath that is connected to a circulation chiller. Two separate chillers were found to provide both the required thermal stability and the necessary cooling power to achieve the desired cooling/warming rates (to be discussed): a NESLAB RTE7 (Thermo Scientific,

2009) and a Julabo FP50-HL (Julabo, 2009). Both circulators come with an integrated tunable/self-tuning PID controller. The controller settings, which will vary slightly based on ambient temperature and heat transfer fluid (a 50:50 volumetric mix of ethylene glycol/water is recommended) should be adjusted by qualified personnel to achieve the requisite temperature stability with no thermal overshoot inside the autoclave. Standard polyethylene foam pipe insulation should be applied to all lines carrying refrigerant to minimize the cooling load imposed on the system and also to decrease the thermal fluctuations at or near the desired setpoint.

6.3.4 Test Procedures and Data Interpretation

Because the exact testing procedures for KHI quality assurance and quality control are device specific (e.g., valve operation, pump operation, computer control, etc.), the guidance given herein will be general for all autoclave-type systems with configurations *similar* to those described above. Where applicable, auxiliary equipment to the autoclave, such as vacuum pumps and high-pressure liquid-charging pumps, will be noted. Safety protocols and appropriate monitoring equipment must always be used when performing these tests, especially when high-pressure, H_2S-containing fluids are used. However, the description of all safety devices attached or pertaining to operation of this test is beyond the scope of this work.

6.3.4.1 T^{eq} Tests

The first step in performing a KHI hold-time test is determining the hydrate equilibrium temperature (T^{eq}) at a desired testing pressure. The reason for this is that KHI performance is typically cited in terms of subcooling, or the amount by which the system can be cooled for a time below T^{eq}. The equation for subcooling is shown in Equation [6.1]:

$$\text{Subcooling} = T^{eq} - T^{actual} \qquad [6.1]$$

While reasonable predictions of T^{eq} can be made (see Figure 6.8), the general accuracy of such measurements is on the order of 1.7°F, an error that can cause several orders-of-magnitude discrepancies in hold time. The most reliable method to determine T^{eq} is, of course, experimentation. The easiest method, experimentally, of obtaining T^{eq} for a specific pressure is to perform an isobaric (constant pressure) test. This was the method employed at Lab B. However, an isobaric test requires the use of a high pressure pump that can often make such a test more complex in terms of equipment, setup, and cost. This test is performed in the following manner:

1. 100 ml of tap water is added to the 300-ml autoclave under atmospheric conditions.

2. 100 ml of stabilized condensate is added to the autoclave under atmospheric conditions.

3. The autoclave is sealed, and the pump used to maintain a constant system pressure. A Teledyne ISCO 260D or 500D high-pressure syringe pump is connected to the autoclave head. If possible, the pump fluid chamber should be insulated with common polyethylene pipe insulation to ensure that ambient fluctuations in temperature do not cause changes in pump volume.

4. The autoclave body, charging lines, and pump lines are evacuated via the use of a vacuum pump, Precision Scientific model DD-90, for 2 min.

5. The vacuum pump is turned off, and the recirculation chiller is set to maintain a temperature of 46°C (115°F).

6. Stirring at 600 rpm is initiated.

7. Process gas from a gas cylinder is introduced to the system through a port in the autoclave head.

   a. The constant-pressure pump should begin withdrawing gas to maintain a constant pressure setpoint (15.5 MPa or 2250 psia) used in all experiments.

   b. Gas should be added until such time that the pump reaches 80% of its capacity (400 ml).

   c. The charging process should be done such that the system is filled with gas over the course of 30 min.

8. The gas charging line from the main gas cylinder to the autoclave cell should be sealed (valved off).

9. The system should be cooled to 35°C (95°F) at a rate of 5.56°C (10°F) per hour.

   a. The aforementioned circulation chillers have the capacity to follow user-programmed quench/heat recipes.

   b. Internal pump volume will decrease due to increased gas solubility and density at lower temperatures.

   c. To ensure water and condensate saturation with vapor, the system should be held at 80°F for at least 30 min, or until such time that the internal pump volume ceases to decrease.

10. The system should be cooled to at least 12.7°C (55°F) at a rate of 3.6°C (6°F) per hour.

    a. Gas volume inside the pump will decrease precipitously upon for-
    mation of hydrate (see Figure 6.10).

11. The system should be warmed to 23.9°C (75°F) at a rate of 0.5°C
    (1°F) per hour.

    a. Slow heating rates are required in this system to ensure that accu-
    rate thermodynamics are not masked by heat transfer effects.

    b. Some studies (Rovetto et al., 2006) have shown heating rates as
    low as 0.12°C (0.22°F) per hour are required, but ExxonMobil
    testing such as shown in Figure 6.13 has not found any discrep-
    ancy between autoclave and miniloop data while passing through
    T^{eq} at a heating rate of 0.5°C (1°F) per hour.

12. The system should be warmed to 46°C (115°F) at a rate of 5.56°C
    (10°F) per hour.

Steps 1 through 12 constitute one cycle of a T^{eq} run. A typical pressure, temperature, and pump volume trace with time is shown in
Figure 6.10.

It can be seen that at a time of approximately 6.8 hr, two events
corresponding to a hydrate event occur. The first, and most dramatic,
is a steep drop in the pump volume. In a constant pressure system, the
pump volume drops as gas is consumed. However, as shown from a time

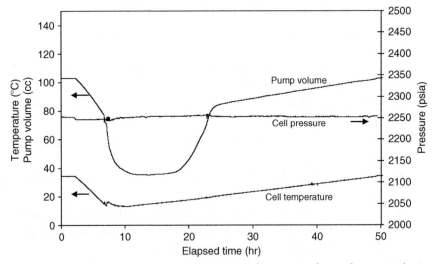

Figure 6.10 System pressure, temperature, and pump volume for an isobaric
T^{eq}-determining test.

of 2.5 to 6.8 hr (where hydrate was not forming), simple solubility and density effects cause the pump volume to drop in a linear fashion with temperature. From 6.8 to 15 hr, however, the volume drop is much more precipitous. At this point, gas and water are forming hydrate. The second required coincident event that signals hydrate formation is a spike in the temperature due to the exothermic formation of hydrate. The rate of hydrate formation in the system overwhelms the cooling rate of the chiller, and therefore the temperature rises. Shortly thereafter, however, the cooling rate increases (due to PID control on the chiller) and the process returns to the desired setpoint. In a system with no metastability (i.e., a system where hydrate forms immediately upon dropping below T^{eq}), the determination of T^{eq} could be made after 10 hr of experimentation. However, hydrates, even without the presence of a KHI, can withstand several degrees of subcooling for a short time prior to formation. The consequence of such a phenomenon is that the temperature spike and pressure drop seen in Figure 6.10 typically occur at a temperature 2 to 6°C *below* the actual hydrate formation temperature. This is the reason for the slow warmup; aside from a few rare cases, hydrate dissociation upon careful warming will not encounter any metastability issues, and is therefore a more reliable means by which T^{eq} can be determined. This is more easily seen by examination of Figure 6.11, a temperature versus pump volume plot from the experiment shown in Figure 6.10.

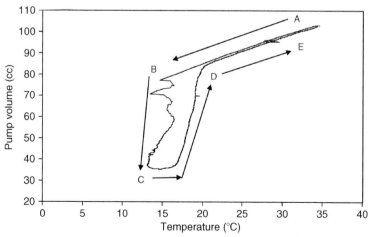

Figure 6.11 Pump volume (cm^3) versus temperature (°C) in a T^{eq} run.

Point A corresponds to the starting point in the experiment, Step 9 from the procedures listed above. As per protocol, Step 10, the system is cooled to 12°C. However, in doing so it passes through the hydrate equilibrium temperature and reaches a point, B, at which the system is metastable and can no longer be cooled without forming hydrate. At this point, the system begins forming hydrate and the pump volume drops precipitously as gas is consumed. Moving from A to B, it can be seen that the pump volume decrease is linear with temperature, and gas was "consumed" by simple contraction and solubility effects. As the system moves from B to C, the hydrate formation exotherm raises the internal temperature of the cell toward the true hydrate equilibrium temperature, but is counteracted by the effect of the chiller. At point C, the system has finished forming hydrate, as indicated by a lack of gas consumption (pump volume decrease). According to Step 11, the system is then warmed at a slow rate from C to E. As it does so, some of the hydrate within the system dissociates and begins to give off gas, causing the pump volume to increase. An important point about the dissociation is that the "true" hydrate equilibrium temperature for this system occurs at point D, when the last crystal of hydrate melts and the system's chemical composition returns to a pre-hydrate state. Prior to this time, at a point between C and D, dissociation *is* occurring but the nonhydrated chemical composition of the system is not equivalent to the as-charged composition, and therefore represents a "false" T^{eq}. Passing point D, the last crystal of hydrate is melted and the system resumes a typical volume versus temperature behavior for a gas in contact with water and condensate. At point E, the system has completed a full thermal cycle, and the pump volume at E should correspond to that observed at A, within the accuracy on the measurement. Determination of T^{eq} is made by examining the point at which the slope changes on the C-to-E warmup curve. From T^{eq} to point E, the slope of the volume versus temperature curve should be nearly identical to that of point A to point B. For statistical validity, the thermal cycle of Steps 9 through 12 should be repeated, at minimum, three times. T^{eq} is then obtained from an average of the three separate tests. Some fluctuation in T^{eq} is expected due to error in the measurement, as shown in Figure 6.12, an overlay of three tests with the T^{eq} point expanded. A standard deviation of 0.2°C (0.36°F) is considered adequate. In this case, the average T^{eq} was found to be 20.68 ± 0.18 °C (69.2°F), which is equivalent to miniloop-determined T^{eq} for the gas A composition.

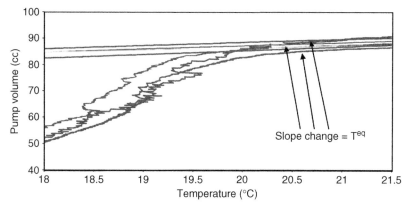

Figure 6.12 Three isobaric T^{eq}-determining tests.

Isochoric or constant volume tests are another means by which autoclaves can be used to determine the hydrate equilibrium temperature. They are as accurate as isobaric tests, but require less equipment, thus making them more robust in terms of maintenance and less prone to failure. However, this simplicity comes at the expense of slightly more experimental effort to match a specific testing pressure. An isochoric test was the method of choice at Lab A, as it lacked a high-pressure injection pump. The isochoric test is run as follows:

1. 100 ml of tap water is added to the 300 ml autoclave under atmospheric conditions.
2. 100 ml of stabilized condensate is added to the autoclave under atmospheric conditions.
3. The autoclave is sealed, and gas charging lines are connected.
4. The autoclave body, charging lines, and pump lines are evacuated via the use of a vacuum pump, Precision Scientific model DD-90, for 2 min.
5. The vacuum pump is turned off, and the recirculation chiller is set to maintain a temperature of 46°C (115°F).
6. Stirring at 600 rpm is initiated.
7. Process gas from a gas cylinder is introduced to the system through a port in the autoclave head. The charging process should be done such that the system is filled with gas over the course of 30 min to allow complete saturation of the process fluids.

Some discussion is warranted here as this step, combined with the desire to match a pressure of 15.5 MPa (2250 psia), takes some effort. Charging is done at a high temperature both to prevent gas hydrates while charging, and for safety to ensure that the maximum pressure experienced by the system will never exceed that of the initial charge. Because the pressure will drop as the system is cooled, it is then desirable to slightly overcharge the system at a higher temperature to match a testing pressure at lower temperatures. The amount of overcharge necessary will change depending on the composition of the fluids used, but to a good approximation, it was found for the fluids in question used in the ratios specified, a linear pressure drop of 92.5 kPa/$^\circ$C, or 7.45 psi/$^\circ$F was experienced. Thus, the system should be charged to a pressure of 17.87 MPa (2592 psia) prior to cooling. Note, however, that KHI hold times are given in terms of subcooling—the temperature at which the test is run compared to the equilibrium temperature for such a condition. Thus, even if the desired pressure is not matched precisely, a hold–time versus subcooling curve may still be generated with a high degree of confidence because T^{eq} will be well known.

8. The gas charging line from the main gas cylinder to the autoclave cell should be sealed (valved off).
9. The system should be cooled to 35°C (95°F) at a rate of 5.56°C (10 $^\circ$F) per hour.
   a. The aforementioned circulation chillers have the capacity to follow user-programmed quench/heat recipes.
   b. System pressure will decrease due to increased gas solubility and density at lower temperatures.
10. The system should be cooled to at least 12.7°C (55°F) at a rate of 3.6°C (6°F) per hour.
    a. System pressure will decrease precipitously upon formation of hydrate (see Figure 6.13).
11. The system should be warmed to 23.9°C (75°F) at a rate of 0.5°C (1°F) per hour.
    a. Slow heating rates are required in this system to ensure that accurate thermodynamics are not masked by heat transfer effects.
    b. Some studies (Rovetto et al., 2006) have shown that heating rates as low as 0.12°C (0.22°F) per hour are required, but ExxonMobil testing such as shown in Figure 6.13 has not found any discrepancy between autoclave and miniloop data while passing through T^{eq} at a heating rate of 0.5°C (1°F) per hour.

12. The system should be warmed to 46°C (115°F) at a rate of 5.56°C (10°F) per hour.

Interpretation of isochoric T^{eq} tests is done in much the same manner as isobaric tests, except that system pressure is observed to drop, rather than a drop in pump volume. Figure 6.13 shows a full T^{eq}-determining experiment performed isochorically using gas B. Cooling of the system was done down to 4°C to demonstrate that complete conversion to hydrate had taken place—this is evidenced by the system resuming a more gradual pressure versus temperature slope below 14°C. Also note that point D, the T^{eq} point, occurs at a slightly lower pressure (14.6 MPa, 2120psi) than that shown in Figure 6.10. Its value, 20.7°C, was essentially identical to that determined for the isobaric test (gas A), however. This allowed for direct and easy comparison of test temperature and subcooling between the two tests, despite the fact that they contain two different chemical compositions.

Figure 6.14 is an expanded view of the pressure–temperature trace shown in Figure 6.13. Similar to Figure 6.12, a clear slope break is observed at the final moment of hydrate crystal dissociation. The same statistical criteria as isobaric tests should be followed in isochoric tests for determining T^{eq}.

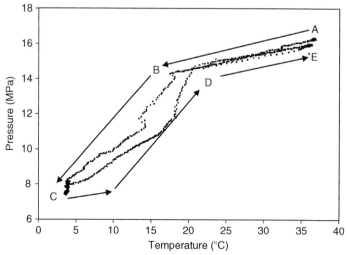

Figure 6.13 System pressure (MPa) versus temperature (°C) in an isochoric T^{eq} run.

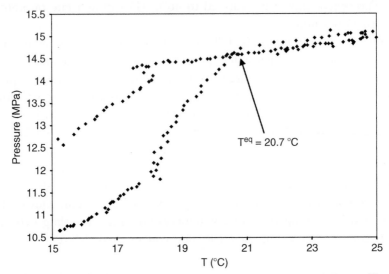

Figure 6.14 System pressure (MPa) versus temperature (°C) in an isochoric T^{eq} run.

6.4 HOLD-TIME TESTS

After T^{eq} has been determined, hold-time studies may commence. Ideally, hold-time tests should be done on exactly the same fluids used in the T^{eq} tests. That is, any chemical injection of KHI or corrosion inhibitor (CI) should be done under pressure, on the same fluids from the T^{eq} tests. Additionally, due to the small volume of KHI and CI typically used in these tests, any injected chemical should be followed with 1 to 2 cm^3 of water to ensure that the full dose of chemical is added to the test; dosage of KHI and CI should be adjusted to account for this additional water. Finally, the injection line diameter and length should be kept as small as possible to prevent excessive holdup of the injected fluids. In the event that chemicals cannot be added to the system under pressure, it is suitable, as a last choice, to premix the KHI, CI, and water under atmospheric conditions prior to adding condensate and pressurizing the cell.

As with the T^{eq} tests, exact test procedures are device specific, and thus general guidance will be given. Hold time tests with high-pressure liquid injection capabilities are performed in the following way:

1. An appropriate dose- and chemical-specific volume of KHI is injected into the autoclave, which should be at a temperature of 46°C (115°F) from Step 12 in the T^{eq} tests.

   a. This temperature, held for 30 min, is thought to eliminate any partially formed hydrate cages that cause KHI tests to fail prematurely.

   b. If this temperature presents physical limits on the equipment, a temperature of 35°C (95°F) may be held for a minimum of 3 hr prior to starting hold-time tests.

2. 2 ml of tap water is injected through the chemical injection lines to wash any residual KHI out of the line.

3. The system is cooled at approximately 0.32°C (0.57°F) per minute to the desired testing temperature.

   a. Although the test temperature will depend on the dosage strength and the chemical to be tested, it is suggested that subcoolings be tested three times each.

   b. Excessively warm temperatures (those approaching T^{eq}) will give hold times on the order of months, which is unsuitably long.

   c. Excessively cold temperatures will give hold times on the order of hours, which approaches the error on the hold-time measurement repeatability; shown to be on the order of 10 hr.

   d. The circulation chiller should be tuned such that no overshoot in cooling is experienced upon reaching the setpoint, as an overshoot of 1°C can cause the test to fail at least an order of magnitude faster than without overshoot.

   e. The circulation chiller should also be tuned such that the setpoint can be reached without an exponential decay to the desired temperature, as any time spent below T^{eq} counts against the hold time.

4. The system is held at the setpoint until hydrate failure is observed (Figure 6.15).

5. The system is then warmed to 35 or 46°C (95 or 115°F) at a rate of 0.32°C (0.57°F) per minute and held for the prescribed time.

As seen in Figure 6.9 and also in Figures 6.15 and 6.16, KHI failure and coincident hydrate formation in an autoclave are marked by the simultaneous temperature rise and pressure/volume drop in the system. These are the only acceptable indicators of hydrate formation in a blind cell. Motor current draw, stirrer torque, and fluid viscosity all typically increase with hydrate formation, but these factors are not as reliably measured in a high-pressure cell.

For statistical validity, hold-time tests should be repeated at least three times per hold temperature. Hydrate nucleation is a statistical process, and thus even in a well-controlled environment, some scatter in the data is expected—typical scatter in the autoclave data was on the order of 10 hr.

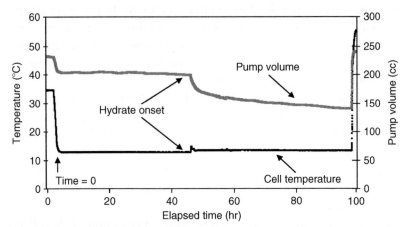

Figure 6.15 Observed hydrate formation via a temperature spike and volume drop in an isobaric autoclave.

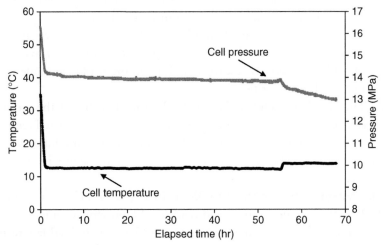

Figure 6.16 Observed hydrate formation via a temperature spike and pressure drop in an isochoric autoclave.

Figure 6.17 shows the results of all hold–time tests at both Laboratories A and B using customary English units on subcooling. Table 6.2 is a summary of experimental conditions in both tests. Also shown in Figure 6.17 is the hold–time curve versus subcooling from the ExxonMobil miniloop. It is important to understand first that the curve shown is a regression of experimental data, and thus actual data is expected to fall both to the left and right within the error on the measurement. Second, although at high

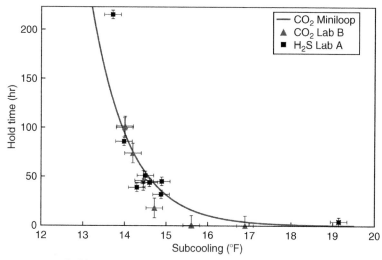

Figure 6.17 KHI hold-time results in a CO_2- and H_2S-containing autoclave compared to CO_2-containing miniloop results.

Table 6.2 Experimental Conditions Used at Laboratories A and B for KHI Hold-time Tests in Autoclaves

Factor	Lab A	Lab B
Gas composition	Gas B	Gas A
Water volume	100 ml	100 ml
Condensate volume	100 ml	100 ml
KHI	KHI A	KHI B
CI	None	CI A

subcoolings (16°F and higher) the curve displays well-behaved exponential decay, the error on hold times at these temperatures approaches that of the measurement. That is, the "signal-to-noise" ratio is 1 or lower. Therefore, more credence and experimental effort should be given to those subcoolings that result in hold times of greater than 24 hr; longer tests (to a point) are better. Finally, the curve is applicable to both the KHI A and KHI B under the conditions shown in Table 6.2, with an understanding that none of the ExxonMobil miniloop tests were performed in the presence of H_2S.

Note that gas A was used by Lab B due to material and safety constraints.

Although there are a few data points that do not fall within 1 standard deviation (shown) of the hold-time curve, all acquired data fall within 2 standard deviations of the curve in either the x- or y-direction. As shown, the error in the temperature measurement plays the largest role in hold times and, as mentioned, should be controlled most tightly.

From Figure 6.17 it can be seen that data for both the H_2S- and CO_2-containing systems agree both with each other and the hold-time curve generated by the ExxonMobil miniloops.

6.4.1 Autoclave Study Summary

It was found that a 300-ml vessel, filled with 100 ml of water, 100 ml of gas condensate, and 100 ml of vapor, stirred at a rate of no less than 600 rpm, and controlled to within $\sim 0.2°C$ of the desired testing temperature successfully reproduced miniloop KHI behavior to within the experimental precision of the miniloop.

It was shown that in the event that KHI testing with hydrogen sulfide (H_2S) cannot be performed due to laboratory or materials constraints, substitution of H_2S with carbon dioxide, another hydrate forming gas, may be determined to give similar kinetic hydrate inhibitor hold times for the same subcooling.

Although efforts were made to demonstrate that KHI QA/QC work in an autoclave can be performed to the standards of a flowloop, the variety of oilfield chemicals and fluid compositions precludes the statement that "autoclaves *always* yield the same results as flowloops do." Using a KHI alone, as well as a KHI/CI pair showed no statistically significant difference from miniloop performance for each case.

Results of this work should not be extrapolated to different geometries or reactor setups.

6.5 STUDY 3: CORRELATION OF MINILOOP, LARGE LOOP, AND ROCKING CELL RESULTS

A study (Talley, Mitchell, and Oelfke, 1999) was conducted to determine the hydrate onset subcooling of several KHIs of differing chemical structure in a 0.5-in miniloop, ½-in rocking cells and a 4-in flowloop. The tests were all carried out at the same cooling rate of 6°F per hour. The KHI concentrations were all 0.5 weight % active polymer. The fluids were all synthetic seawater, with King Ranch condensate and synthetic Green

Canyon gas mixture at 1000 psig. The hydrate equilibrium dissociation temperature was 64°F ± 0.5°F for all tests.

The subcoolings in the miniloop ranged from 10 to 30°F and the subcoolings in the 4-in flowloop ranged from 10 to 28°F for the same KHI polymers. The individual structures were observed to have subcoolings that correlated to within 2°F between the 0.5-in miniloop and the 4-in flowloop.

The subcoolings in the miniloop ranged from 10 to 30°F and the subcoolings in the sapphire rocking cell ranged from 17 to 24°F for the same KHI polymers. The high and low ends of subcooling for individual KHI structures correlated only to within 6 to 7°F between the 0.5-in miniloop and the sapphire rocking cells. The rocking cell results have a distinct correlation to miniloop results that differs from the 4-in flowloop correlation.

Miniloop and 4-in flowloop results (reported in Talley, Mitchell, and Oelfke, 1999) are only representative of turbulent, flowing tests. If the flow regime in any of the loops is stratified flow or wavy flow, the onset subcoolings will be unpredictably different, usually trending toward less subcooling.

6.6 STUDY 4—CORRELATION OF LARGE LOOP AND FIELD RESULTS

A commercial KHI was tested in a turbulent, flowing 4-in flowloop and in an 8-in export pipeline offshore under the same conditions of composition, pressure, flow regime, and temperature (Talley and Mitchell, 1999). In each case, the KHI dose rate was decreased until hydrates formed within hours of reaching a certain fluid temperature.

The field study was conducted on the South Pass 89A platform (see Figure 6.18). The KHI-treated pipeline was an 8-in wet gas export line in 450 ft of water in the Gulf of Mexico.

The pipeline flowed at 15 to 18 MMSCFD sweet gas with 3 bpd of condensed water. The inlet pressure and temperature were 1230 psig and 110°F, respectively. The cold-point temperature range was 53 to 60°F and the hydrate equilibrium temperature was 66 °F. The subcooling was 6 to 13°F year round.

The field demonstration used a commercial KHI based on ExxonMobil's patented VIMA-VCap chemistry (see Figure 5.4 in Chapter 5 for chemical structure). KHI injection was decreased into the

Figure 6.18 South Pass 89A in the Gulf of Mexico.

export line once per day until the line began to hydrate. This was done at a subcooling of 6°F. The highest KHI concentration that failed was 0.05%.

The KHI concentration that fails after 1 day in the large flowloop is about 0.04% at 6°F subcooling. Given the uncertainties in condensed water volume in the pipeline, the agreement between field and flowloop is excellent.

6.7 CONCLUSION

The main questions addressed in this chapter are what test apparatus sizes and what test apparatus flow regimes adequately predict field performance. The studies cited here have shown that results in flowloops as small as 0.5-in diameter agree with results from 4-in flowloops. It seems reasonable that all sizes in between will also agree, subject to the conditions suggested in these studies. These studies have also shown that results from 0.5-in and 4-in flowloops agree with results in 8-in field pipelines.

Agreement was achieved for turbulent, flowing systems. Shut-in conditions, stratified flow or wavy flow, did not agree with turbulent flow results. We conclude that a broad range of lab test apparatuses can predict field performance provided that the flow characteristics, fluid compositions, pressures, and temperatures can be simulated.

REFERENCES

Study 1

Klomp, U.C., 2008. The World of LDHI: From Conception to Development to Implementation, Proceedings of the 6th International Conference on Gas Hydrates, Vancouver, BC, Canada, July 6–10.

Talley, L.D., 2003. Relationship between Hydrate-Free Hold Time and Subcooling for a Natural Gas System Containing Kinetic Hydrate Inhibitors, Nucleation Workshop at Ecole Nationale Supérieure des Mines de Saint Etienne, June 17–19.

Study 2

Anderson, G., 2002. Solubility of carbon dioxide in water under incipient clathrate formation conditions. *Journal of Chemical and Engineering Data* 47, 219–222.

Autoclave Engineers, accessed March 2, 2009. Agitator/Mixer Product Comparison. http://www.autoclaveengineers.com/ae_pdfs/AG_Agitator_ mixer_comparison.pdf.

Autoclave Engineers, 2009a. 100 & 300 ml Bolted Closure Stirred Laboratory Reactor. http://www.autoclaveengineers.com/ae_pdfs/SR_100_300_ BoltedClosure_Tech.pdf.

Englezos, P., Dholabhai, P., Kalogerakis, N., Bishnoi, P.R., 1987. Kinetics of formation of methane and ethane gas hydrates. *Chemical Engineering Science* 42, 2647–2658.

Fleyfel, F., Song, K.Y., Kook, A., Martin, R., Kobayashi, R., 1993. Interpretation of carbon-13 NMR of methane/propane hydrates in the metastable/nonequilibrium region. *J. Phys. Chem.* 97 (25), 6722–6725.

Herri, J., Gruy, F., Cournil, M., 1996. Kinetics of methane hydrate formation. Proceedings of the 2nd International Conference on Natural Gas Hydrates, Toulouse, France, June 2 6, pp. 243–250.

Julabo, accessed March 2, 2009. FP50-HL Refrigerated/Heating Circulator. http://www.julabo.com/us/p_datasheet.asp?Produkt=FP50-HL.

Lederhos, J.P., Long, J.P., Sum, A.K., Christiansen, R.L., Sloan, E.D., 1996. Effective Kinetic Inhibitors for Natural Gas Hydrates, *Chemical Engineering Science* 51 (8), 1221–1229.

Rovetto, L.J., Strobel, T.A., Koh, C.A., Sloan Jr., E.D., 2006. *Fluid Phase Equilibria* 247 (1–2), 84–89.

Sloan, E.D., Koh, C.A., 2008. *Clathrate Hydrates of Natural Gases*, third ed. Taylor & Francis, CRC Press, Boca Raton, FL.

Thermo Scientific, accessed March 2, 2009. Neslab RTE Digital Plus Refrigerated Circulators. http://www.thermoscientific.com/wps/portal/ts/products/detail?navigationId= L10544&categoryId=87419&productId=11962717.

Turner, D., 2005. Clathrate Hydrate Formation in Water-in-oil Dispersions. Chemical Engineering. Ph. D Thesis. Colorado School of Mines, Golden.

Study 3

Talley, L.D., Mitchell, G.F., Oelfke, R.H., 1999. Comparison of Lab Results on Hydrate Induction Rates in a THF Rig, High-Pressure Rocking Cell, Miniloop and Large Flowloop, Proceedings of the 3rd International Conference on Gas Hydrates (ICGH 1999), Salt Lake City, Utah, July 18–22.

Study 4

Talley, L.D., Mitchell, G.F., 1999. Application of proprietary kinetic hydrate inhibitors in gas flowlines, 31st Annual SPE Offshore Technology Conf. (Houston, 5/3–6/1999) PRO3. 3, pp. 681–689. (OTC-11036).

Appendix
Reed, R.L., Kelley, L.R., Neumann, D.L., Oelfke, R.H., Young, W.D., 1994. Some
   preliminary results from a pilot-size hydrate flowloop. *Ann. N.Y. Acad. Sci.* 715, 311.

APPENDIX
High-Pressure Miniloop Procedures
Facility Description

The miniloop studies presented in this chapter were conducted in high-
pressure miniloops (HPML-1 and HPML-2), which are 0.5-in diameter,
32-ft loops. A schematic of HPML-1 is given in Figure 6.19. The other
miniloop is essentially the same as HPML-1.

The overall design of ExxonMobil miniloops is similar to that of the 4-in
diameter ExxonMobil flowloop (Reed et al., 1994). HPML-1 and HPML-2
are both closed loops built of heavy wall tubing through which a mixture of
pressurized hydrocarbon gases, liquid hydrocarbons, and brine are circulated.
A positive displacement gear pump is used in HPML-1 experiments and a
sliding vane pump is used in HPML-2 experiments. Both types of pumps
circulate multiphase mixtures through the loop to simulate pipe flow and
to keep the system well mixed. HPML-1 operates at pressures up to 5000
psig. HPML-2 operates at pressures up to 10,000 psig.

The miniloops normally operate at constant pressure. HPML-1 is
connected to a 3-liter gas accumulator. The total volume of tubing plus
gas accumulator is 4619 cc (1659 cc in the wetted loop). HPML-2 is
connected to a 3-liter gas accumulator with a piston spacer that limits
the internal volume. The total volume of tubing plus gas accumulator in
HPML-2 is 3565 cc (1655 cc in the wetted loop). A pressure-controlled
hydraulic oil system maintains constant pressure against a floating piston
inside the gas accumulator to maintain constant gas pressure in the loop.
The accumulators of HPML-1 and HPML-2 are maintained at nearly constant temperature settings selected between 85 and 115°F. The liquids are
not circulated through the accumulator but the gases are.

The internally wetted loop tubing is immersed in a temperature-
controlled glycol/water bath. Automated heating and cooling systems
ramp the loop temperature up and down at predetermined rates. The
loops typically operate at temperatures between 30 and 125°F. The temperature is controlled within ±0.3°F of the setpoint. This is one of the
more important features of the mini-loops as an error in 1°F can result
in an order of magnitude difference in KHI hold times.

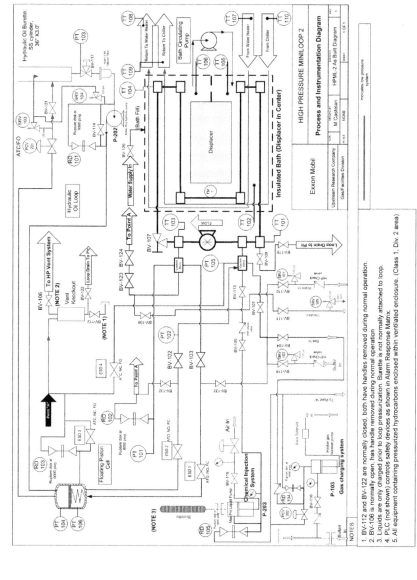

Figure 6.19 Process and instrumentation diagram for high-pressure miniloop 2.

Experimental Fluids

Stabilized liquid condensate and synthetic gas mixtures were combined in the miniloop to simulate reservoir hydrocarbon fluids. Tap water was combined in the miniloop to simulate a condensed water composition.

Miniloop Procedures

The miniloop procedures have been modified to optimize conditions for measuring hydrate nucleation. The following conditions must be met to ensure accurate and reproducible results:

- Must be able to accurately measure the hydrate dissociation temperature.
- Composition such that KHI is chemically and biologically stable.
- Driving force must be constant/reproducible.
  - Temperature is maintained uniform (less than 0.3°F variation).
  - Pressure must be held constant.
  - Excess of water and/or guest so that composition remains constant.
  - The loop circulation pump speed and volumetric pump rate were kept constant throughout the study unless specifically noted.
- Surface area needs to be large and continually renewed by shear/mixing.
- Water structure is initialized by heating the system above the equilibrium temperature at which the memory effect is erased.
- The gas in the gas reservoir was continually exchanged with gas in the loop to maintain nearly constant composition during T_{eq} experiments.
- Early detection of hydrates insured low conversion of water and hydrate-forming gas components in KHI experiments. Low conversion of water and gas to hydrates insured that gas compositions did not change significantly before hydrate detection occurred.
- To ensure that these conditions are met, the following procedures have been developed to test KHI performance.

Loop Charging

The loop was cleaned using a sequence of fluids including gas condensate and clear water. The loop was then blown dry with nitrogen before charging. Periodically, the loop is cleaned further by adding bleach to the system reducing bacterial growth, which is antagonistic to KHI performance. A vacuum was pulled on the loop and the accumulator with the accumulator piston at the top of the accumulator (maximum gas volume).

Gas was charged to the loop through the accumulator with the temperatures of the accumulator and loop set at 85 to 115°F. The gas temperature also needs to be injected at 75°F to ensure consistent compositions. The charge pressure was always lower than the desired experimental loop pressure so that the accumulator piston would always be free floating below the top end cap of the accumulator while at experimental pressure. The condensate and brine were then added at room temperature under pressure. The loop pressure was then put under computer control and the loop was ready to begin the experimental sequence.

Hydrate Detection

Hydrate formation in the miniloops was normally detected by observing a change in the differential pressure across the circulation pump. The observations that help to determine that hydrate formation is occurring are follow:

1. Detection of an increased differential pressure drop across the circulation pump in a flowing system
2. Detection of a decreased gas accumulator volume in excess of vapor-liquid thermal equilibration
3. Detection of a rapid decrease in loop pressure in excess of vapor-liquid thermal equilibration
4. Detection of an increase in loop process temperature due to exothermic hydrate heat of formation
5. Detection of an increased temperature in the gas transfer line from the wetted loop to the accumulator due to loss of gas flow
6. Detection of differential pressure transient oscillations
7. Single pressure spikes across the circulation pump
8. Visual observation of hydrate solids

We define hydrate onset temperatures based on hydrate detection by any of the above mentioned methods. Hydrate formation can be detected in each miniloop apparatus by more than one of the observations.

The beginning of hydrate nucleation starts somewhere between when the system drops below the hydrate equilibrium temperature and when hydrates are first detected. The ideal procedure to measure nucleation time is to drop the process temperature instantaneously from a temperature above the equilibrium temperature to the hold temperature. In practice, the process temperature is subcooled over a period of an hour. Hold times less than a few hours are not accurate measures of nucleation time at fixed temperature.

T_{onset} is defined as the temperature of a system being cooled at which hydrates are first detected. It is known that invisible hydrate nucleation and crystal growth events actually take place before a system reaches T_{onset}.

Measurement of Hydrate Equilibrium Temperature in Miniloops

The hydrate equilibrium temperature must be determined each time a miniloop is charged with fluids to verify the charge was done properly. The experimental equilibrium temperatures vary less than 0.3°F for multiple charges of the same fluids when done properly.

Hydrate disappearance temperature is commonly regarded as the closest approximation to the hydrate equilibrium temperature, T_{eq}. The following methods are commonly used to determine T_{eq} in miniloops:

1. Gas volume changes at constant pressure
2. Gas pressure changes at constant volume
3. Differential pressure spikes across an element in the flowing system
4. Visual observation of hydrate melting
5. Homogeneous fluid temperature in the flowing system

In this study, the experimental T_{eq} was defined as the first temperature during warm-up in a hydrated system at which no spikes were detectable in the differential pressure across the circulating pump due to hydrate particles. This generally occurs when the volume returns to the same level as before hydrate formation. Of course, hydrate dissociation occurs before this point and so the miniloop measures when the last quantity of hydrate leaves the system.

When measuring T_{eq}, caution must be exercised in multicomponent experiments that the observed hydrate decomposition occurs at constant water, hydrate, and gas phase compositions. If the salinity, gas composition, or hydrate type were to change after the onset of hydrate formation, the measured value of T_{eq} would be shifted relative to the actual value for the initial compositions. Experimentally, if the water and gas fractions in hydrates are small, then the compositions of the bulk gas and water phases will be constant during the measurement of T_{eq}.

In this study, we follow the convention that subcooling is defined as the difference between the hydrate equilibrium temperature and the experimental temperature, $T_{eq} - T_{exp}$. The maximum subcooling achievable in a lab experiment is $T_{eq} - T_{onset}$. For this reason, the last indication of hydrate present in the system is taken at the hydrate equilibrium as this temperature marks the maximum subcooling that the KHI will need to be tested against. This measurement of hydrate equilibrium gives the most conservative results in regards to KHI performance in the field.

In this study, hydrate dissociation was carried out over a sufficiently long time that the heating rate no longer affected the measured temperature at which hydrates disappeared. Experience has shown that a heating rate of $2°F$ per hour is slow enough to determine T_{eq} reproducibly and accurately in a 0.5-in miniloop.

T_{eq} was measured before adding kinetic hydrate inhibitor or other chemicals to a system. The presence of kinetic inhibitor causes the system to equilibrate slowly and makes the measurement of T_{eq} difficult.

Miniloop Results versus Rocking Cell Results

Most hydrate analytical labs do not perform flowloop studies of KHIs (or AAs). Most labs obtain performance data in screening tests based on rocking cells, stirred cells and/or stirred tank reactors. Their experiments are usually carried out at constant volume and variable pressure, where pressure drops during cooldown and hydrate formation. They usually use field liquid samples combined with a lab gas mixture. Their results are often difficult to reconcile with results obtained in flowloops. The main reasons have to do with (1) differences of hydrate nucleation rates and (2) differences of hydrate growth rates, both caused by differences in liquid hydraulics.

The driving forces for hydrate nucleation, growth and agglomeration depend on gas/oil/water composition, temperature, heat conduction, pressure, fluid hydrate history, and mixing/flow characteristics. Various types of lab equipment have been devised to simulate the hydrate formation process in the field. These include rocking cells, stirred reactors and flowloops. Simulation of actual field conditions is most nearly achieved in flowloop experiments. This is apparent from the reproducibility of multiple runs in miniloops versus the relatively scattered results in rocking cells.

While it is possible to devise a rocking cell test whose severity is equal to or greater than the severity of a miniloop test, data scatter in rocking cell tests obscures the comparison to miniloop results. Thus, rocking cell testing can detect KHI antagonism of the worst kind. But the antagonism between some KHIs and CIs may be obscured in rocking cell tests for pairs that otherwise are differentiated in miniloop testing.

Experimental Data Processing

Flowing Hold Time versus Subcooling

In theory, a water/natural gas system could be cooled instantaneously to any amount of subcooling. However, in liquid bath-cooled mini-loops and other systems that change temperature at a rate of 10–100°F per hour, only hold times greater than about 10 hr can be measured accurately.

In this study, flowing hold times were measured at subcoolings ranging from 12 to 23°F (at T_{onset}). The function, flowing hold time versus subcooling, was observed to be nonlinear in subcooling for all KHI concentrations.

There are two known trends in the fitted curves. One is an exponential increase in the flowing hold time with decreasing subcooling. The second is that the flowing hold-time curves shift to higher subcooling as the KHI dose rate increases.

Derivation of Chemical Kinetics Equation

Sloan published four figures that show the relationship between hydrate nucleation rates and driving force in uninhibited systems (see Figure 3.17, pp. 102–103, in E.D. Sloan, *Clathrate Hydrates of Natural Gases*, 2nd ed., Marcel Dekker, 1998). The functional form of the rate versus driving force is exponential in driving force, that is,

$$\text{Hydrate nucleation rate} = A_0 \exp(B_0 \Delta G) \qquad [6A.1]$$

where ΔG is the Gibb's free energy term.

In these experiments, the term ΔG is not rigorously known. However, subcooling of inhibited fluids relative to uninhibited fluids is known. If one assumes an approximate exponential relationship of subcooling, ΔT, to hydrate appearance rate, one obtains

$$\text{Hydrate appearance rate} = A \exp(B \Delta T) \qquad [6A.2]$$

where the coefficients, A and B, are determined by empirical data fitting. The hydrate appearance rate is set equal to 1/flowing hold time in units of hours^{-1}.

Substitution of inverse flowing hold time for rate in Equation [6A.2] gives:

$$1/\text{Flowing hold time} = A \exp(B \Delta T) \qquad [6A.3]$$

Taking the natural log of Equation [6A.3] yields:

$$\ln(1/\text{hold time}) = m\Delta T + b \qquad [6A.4]$$

which is linear in subcooling, ΔT, in units of degrees Fahrenheit. The best fit for Equation [6A.4] has been found equivalent for several KHIs where $m = 1.0925$. Values of "b" typically range between -18 and -22, depending on the KHI concentration.

CHAPTER SEVEN

Industrial Operating Procedures for Hydrate Control

Adam Ballard, George Shoup, and Dendy Sloan

Contents

7.1 INTRODUCTION

As noted in Chapter 2 on how plugs form, hydrate blockages are normally successfully avoided; however, when they do form, recommendations

Natural Gas Hydrates in Flow Assurance
ISBN 978-1-85617-945-4
DOI: 10.1016/B978-1-85617-945-4.00007-8

for removal and safety concerns associated with it are detailed in Chapters 3 and 4. It is important to note that hydrate incidents typically occur during transient and abnormal operations such as

1. on startup
2. on restart following an emergency operational shut-in
3. when uninhibited water is present due to dehydrator failure or inhibitor injection failure
4. or when cooling occurs with flow expansion across a restriction.

Hydrate plug formation does not occur during normal flowline operation due to the system design for flow assurance.

Hydrate control during startup, shutdown, and steady-state production operations is maintained by providing proper understanding and guidance to operations, with the following four components:

1. Sound system design
2. Clear documentation of hydrate risks, control methods, monitoring methods, and contingencies
3. Sound operating procedures
4. Trained operations staff

In this chapter the reader will learn the third component of the previous four, namely the generation of an acceptable operating procedure for dealing with hydrates.

Note that an operating procedure is written to cover about 80 to 90% of normal operating conditions; special considerations lying outside the 80 to 90% normal operating conditions, such as hydrate blockage remediation, should be considered as in Chapter 4.

7.2 DEEPWATER SYSTEM DESIGN

Deepwater subsea production oil systems are typically designed as a piggable loop, with active or passive heat management for hydrate control. A looped system allows for:

- Displacement of hydrate prone fluids following a shutdown
- Pigging for sand and corrosion management
- Intelligent pigging for integrity management

Oil systems use inhibitors for startup, and usually run without continuous inhibitor injection after the system is sufficiently warm. The large water cut of oil wells in midlife to late life prevents the use of continuous inhibitor injection. Some single line tiebacks are in operation, but these are relatively few and are usually the result of brown field developments with process and riser constraints.

Deepwater subsea gas systems tend to be bare pipe in all but the shortest tiebacks. Gas systems do not have the initial heat content of oils per unit volume, they produce less water, and they cool due to expansion as pressure decreases in the flowline. For these reasons, gas systems tend to use inhibitors rather than heat to manage hydrates. Liquid control by periodic pigging and integrity management normally mandate dual-line designs for gas.

Highly trained offshore personnel operate both oil and gas fields. Operators typically do not have engineering degrees; however, they do have many years of technical training and many years of experience within the industry. In many cases, those years of operating experience are the best guide to physical realities.

Operators and engineers jointly write the operating procedures for a facility once the design and drawings are mature. The procedures include startup, shutdown, steady-state, and upset conditions. Mitigation plans are identified for hydrate blockage detection and removal, as indicated in Chapter 4.

Integrated field training simulators are routinely in use today. The simulators are "full field models" that include reservoir boundary conditions, wellbores, trees, manifolds, flowlines, risers, boarding chokes, and topsides process equipment, including export pumps and compressors. The subsea and topsides models include the process controllers, sensors, and process and emergency shutdown (PSD and ESD) logic. Such full field models provide a realistic simulation of operations for training purposes.

The same technology of a training simulator can be used to build an online real-time simulator to monitor and forecast events in real time. Real-time simulators are a powerful new tool for bringing flow assurance engineering and best practices into the field.

7.3 APPLICATIONS OF CHAPTERS 1 THROUGH 6

The previous six chapters indicate a significant amount of research and experience regarding hydrates and production operations, some of which is accepted by industry, yet some of which is at the forefront of technology and has yet to be implemented. From a design and operations perspective, it is vital for the flow assurance engineer to consider the following questions:

1. When and where are hydrates likely to form in the production system?
2. What can I control in order to prevent hydrates from forming?

3. What are the monitoring points in the system that will give indication of hydrates?
4. If a hydrate plug forms in the production system, how can it be remediated?

7.3.1 Question 1: When and Where Are Hydrates Likely to Form in the Production System?

Understanding the thermodynamics of hydrates is extremely important to determining where hydrates will form in the production system. Three components, in addition to hydrocarbons, are essential to forming hydrates in hydrocarbon production systems: (1) cold temperature, (2) high pressure, and (3) water. Since all deepwater production systems are conducive to forming hydrates, it is important to understand (via simulation) the likely pressures, temperatures, and water cuts of the system over field life. Some common rules of thumb to keep in mind are:

- If there is no water, then hydrates will not form.
- Gas systems
  When: Most likely to form hydrates during normal and startup operating scenarios.
  Where: Most likely to form hydrates in the subsea system in areas where water collects and/or areas where flow direction is changed. Most likely to form hydrates in the well across the choke.
- Oil systems
  When: Most likely to form hydrates during startup operating scenarios.
  Where: Most likely to form hydrates in the subsea system in areas where gas and/or water have broken out of solution during startup operating scenarios.
- Subsea systems: Most likely to form hydrates near jumpers and/or risers.
- Dry tree wells: Most likely to form hydrates in the riser.

  The typical tools used to understand the question posed are a multiphase thermodynamic package, and a transient, multiphase, thermal–hydraulic package. Key pieces of information to gather are hydrate formation curve and pressures, temperatures, and location of gas, oil, and water over all aspects of production.

7.3.2 Question 2: What Can I Control in Order to Prevent Hydrates from Forming?

In one form or another, all three requirements for hydrate formation can be controlled. Temperature is typically a function of flow rate, but for gas systems in particular, it is also a strong function of where pressure drops occur in the system. Pressure can sometimes be controlled by choking at the wellhead, reducing pressure in the subsea system. Both temperature and pressure management will vary with the field life.

Controlling water is not as simple as temperature and pressure. Water rate and location in the pipe will vary dramatically with field life. Compared to temperature and pressure design predictions over field life, predictions of produced water rates and water cut over field life are relatively unreliable. Therefore, for production operations, the most common method of controlling water for hydrate mitigation is via chemical injection, using the mechanisms in Chapter 5. For gas-dominant systems, hydrate prevention is the method of choice using thermodynamic inhibitors such as methanol and glycol. For oil-dominant systems, a combination of hydrate prevention and hydrate plug prevention is becoming the method of choice. Thermodynamic inhibitors are used to prevent hydrates across restriction (e.g., chokes) during transient operations. For relatively short pipelines, kinetic inhibitors or anti-agglomerants are used during transient operations while, for long pipelines, they are used continuously. The distinction between short and long in this case, is directly related to the temperature of the production fluid in the pipe.

7.3.3 Question 3: What Are the Monitoring Points in the System That Will Give Indication of Hydrates?

As discussed previously, the three key components of hydrates plugging a production system are low temperature, high pressure, and water. Therefore, it is essential in providing the capability to monitor system parameters: temperature transmitters, pressure transmitters, chemical injection, and water monitoring (three-phase separators with water-cut meters). It is not feasible to provide monitoring capability throughout the entire production system. For deepwater systems, it is typical to provide downhole and tree pressure (P) and temperature (T) monitoring capability in wells. For subsea systems, typical P and T monitoring is provided at commingling locations (manifolds) and at the top of incoming risers. Pressures and temperatures between monitoring locations may be calculated.

In a different manner than pressure and temperature, water is usually monitored in the process system using three-phase separation coupled with meters. The water rate of a particular well is known typically only when the well is in test. Between well tests, the water rate of the well is inferred.

Knowing the temperature, pressure, and water rate of the system is required to understand whether (1) the production system is at risk of hydrate plugging and (2) the system is showing signs of hydrate plugging.

7.3.3.1 Risk of Hydrate Plugging

From a conservative standpoint, risk of hydrate plugging is essentially the same as the pressure and temperature conditions being conducive for hydrate formation. As shown in Chapters 2 and 4 of this book, being inside the hydrate formation region does not guarantee hydrate plugging. However, due to the production and safety impacts of a hydrate plug, when possible, the standard approach in deepwater production system design is to prevent hydrate formation altogether, by avoidance of the hydrate pressure and temperature formation conditions. Producing into the hydrate formation pressure–temperature region with unprotected water is not a technique normally considered acceptable in design; however, this technique may be employed as a contingency.

7.3.3.2 Signs of Hydrate Plugging

If a production system enters the hydrate formation region, pressure and temperature monitoring enables operations to take action if indications of a hydrate plug realize. Depending on the number and location of monitoring points, indications of where the hydrate plug is forming can also be determined. In general, if a given pressure transmitter unexpectedly decreases, it is indication that a hydrate plug may be forming upstream of it. If pressure unexpectedly increases, it is indication that a hydrate plug may be forming downstream of it. Figure 7.1 shows the common signs of a hydrate plug forming.

The pressure monitoring areas shown in Table 7.1 are usually the minimum that a subsea development incorporates. If ever a hydrate plug were to form in the system, the precision of plug location, and therefore the method and potential safety implications of remediation, is limited to the number of monitoring locations in the system. Chapter 4 provides more detailed discussion of plug location devices such as gamma ray densitometers and hoop strain gauges; however, pinpoint accuracy of plug location is not necessary.

A secondary method for determining hydrate formation in the production system is to monitor for unexpected increase in fluid temperature. Surrounding temperature increases due to the exothermic nature of hydrate formation. Pressure is the most reliable method for monitoring.

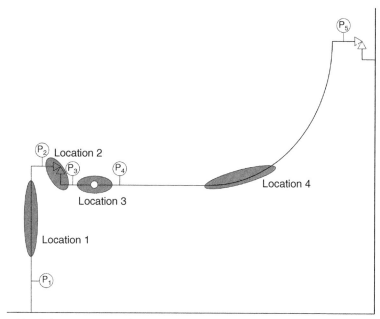

Figure 7.1 Subsea system and potential location of hydrate blockage.

Table 7.1 Pressure Response of Hydrate Blockage Forming

	Location 1	Location 2	Location 3	Location 4
DHP (P_1)	Increase	Increase	Increase	Increase
P_2	Decrease	Increase	Increase	Increase
P_3	Decrease	Decrease	Increase	Increase
SSP (P_4)	Decrease	Decrease	Decrease	Increase
TRP (P_5)	Decrease	Decrease	Decrease	Decrease

DHP, downhole pressure in well; P2, tree pressure upstream of well choke; P3, tree pressure downstream of well choke; SSP, subsea pressure (usually at manifold); TRP, top of riser pressure (upstream of topsides choke).

7.3.4 Question 4: If a Hydrate Plug Forms in the Production System, How Can It Be Remediated?

In this section, we will not go over the many remediation approaches discussed in Chapter 4. Instead, we will discuss, from an operations perspective, how to remediate a hydrate blockage that is thought to be forming. Using the notation for hydrate location shown in Figure 7.1, we will discuss the common schemes for mitigating a complete blockage once it has been identified that hydrates are forming. There are four operations considered: (1) inject chemical, (2) stop flow, (3) reduce flow, and (4) increase flow. Each of these is discussed in the following sections.

7.3.4.1 Inject Chemical

Injection of a thermodynamic inhibitor is almost always the best action to take when there are signs of a hydrate blockage forming. Before injecting inhibitor, it is essential that the location of the blockage has been determined via the pressure response from table 7.1 and the methods of Chapter 4, so that the chemical can be injected closest to the blockage. As well, inhibitor needs to be applied such that it over protects the water in the flow stream. Injecting more chemical than needed provides a chance that hydrates already in the system will be melted.

7.3.4.2 Stop Flow

One of the most natural responses that an operator has when confronted with knowledge that a hydrate plug is forming in the production system is to stop flow. In some instances, stopping flow may be the correct action. If flow is stopped, then an immediate blowdown of the system to a pressure below the hydrate pressure should be performed. Note that blowdown of a gas-dominant system will, in most cases, reduce the temperature in the subsea system potentially making the system more conducive to hydrate formation.

7.3.4.3 Reduce Flow

Reduction of flow alone is usually not going to do anything toward eliminating the hydrate blockage. In fact, temperature in the system will usually reduce as well, pushing the system further into the hydrate formation region. However, reduction of flow as a complement to chemical injection can sometimes be very effective to send an overprotected water phase to the restriction.

7.3.4.4 Increase Flow

Depending on circumstances, increasing flow or "pushing through" could be an appropriate course of action. Increasing flow (thereby increasing temperature), in some instances, will melt any hydrates that are forming in the system. Note that this is not a recommended method for gas-dominated systems.

Understanding the answer to each of the above four questions, for engineers and operators, will lead to safer, more reliable, and more efficient operations. As the operators of the production system are not likely to be heavily involved in the up-front engineering of a deepwater production system, it is important they be informed via operating procedures or guidelines. The key is to include all information that has been gathered to prevent, monitor, and mitigate hydrates in the wells and subsea system. If this information is not presented in an efficient and simple manner via an operating procedure or guideline, then engineering has failed.

7.4 GENERATION OF OPERATING PROCEDURES FOR HYDRATE CONTROL

As discussed previously, clear communication of the hydrate risks to operations is required to produce efficiently. For most deepwater systems, formation of a hydrate plug creates potential for significant safety, production, and environmental impacts. Due to these potential impacts, it is often best to operate conservatively to prevent hydrates in the production system. What follows in this section is a recommended path to communicate hydrate control from design to operations, with the ultimate deliverable of operating procedures.

7.4.1 Detailed Design—Customer : Engineering

Clear documentation of all detailed engineering work should be given in report format to lay out all relevant hydrate information. Examples of this would be fluid compositions, insulation design, hydrate curves, operations no-touch time following shutdown, and thermal hydraulics during all expected operations. Reference to all assumptions and software used for this work should be given. This work is typically generated by the engineering contractor for a project.

7.4.2 Operating Guidelines—Customer : Engineering and Operations

The operating guidelines are meant to act as a general discussion of how the production system is intended to operate. In the context of hydrates, specific sections of these guidelines should lay out the following:
1. Hydrate risk (when and where): Clear documentation of when hydrates are expected to be a concern should be given. Definition

should be given on startup versus steady-state operations; assuming the hydrate control strategy is different according to each mode of operation. The distinction between these two modes of operation (in the context of hydrates) is typically given by temperature somewhere in the system.

2. Hydrate control strategy: For different modes of operation, the hydrate control strategy should be given. For example:

   • During startup operations, chemical injection at location x should be applied until temperature at location y is $> T_H$, the hydrate temperature.

   • During steady-state operations (temperature at location y is $> T_H$), no chemical injection is required. If an unplanned shut-in occurs, operations have z hours until the subsea system is prone to hydrates.

   • During shutdown operations, inject x gallons of chemical to location y and displace subsea system with dead oil.

     For simplicity, it is recommended that a single hydrate temperature (T_H) be used for all documentation in the context of the operating guidelines. The hydrate temperature in the subsea system should not be presented as a curve that is dependent on pressure as this could lead to confusion in the heat of the moment.

3. Monitoring points

4. Contingencies

   The operating guidelines should cover all aspects of flow assurance in production operations; of which hydrates are a major, but only one of several concerns. As such, it is important to provide clear direction as to the major risks in the system and where hydrate concerns occur in that register. Without such direction, an over focus on hydrates may lead to a hydrate-free system that has plugged due to paraffin, scale, asphaltene, sand, and so on.

7.4.3 Operating Procedures—Customer: Operations

In production operations, the operating procedure is typically the only document in hand when executing work; that is to say, the operating guidelines are known by operations but not necessarily consulted. With this in mind, only the high-level points covered in the operating guidelines should be given in the procedure. There will typically not be a specific procedure for hydrate control during an operation. Instead, the steps to control and monitor for hydrates will be embedded into the procedure with specific operations such as:

Step	Remarks
. . .	
5. Open valve CIT2; inject methanol at 5 gpm	Monitor dP for choke blockage
6. . . .	
7. when T2 > 80°F, close CIT2	
. . .	

It is important to note that operation of a deepwater production system is not a simple task and that too much information, especially in an operating procedure, will likely cause confusion and dilute the ultimate goal of the procedure (e.g., startup of a subsea well).

7.5 OPERATING PROCEDURE DETAILS

Consider Figure 7.2, which shows a typical production system, including some of the flow assurance issues:

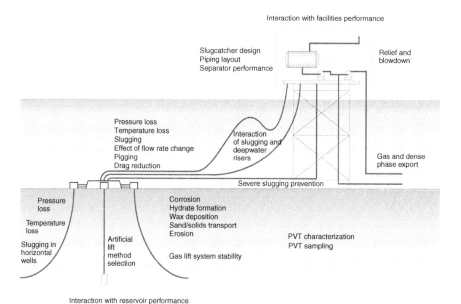

Figure 7.2 A typical production system with some flow assurance problems labeled to indicate the complexity of the entire flow assurance endeavor.

Operating procedures are typically written in modular format for specific operations. Separate procedures should be specified for the following situations:

1. Well startup
2. Well shutdown
3. Well testing
4. Subsea system startup
5. Subsea system shutdown
6. Subsea system blowdown
7. Topsides (or platform) system startup
8. Topsides system shutdown

For a larger-scope operation, a sequence of procedures linking the above modules is normally given. A complete, detailed hydrates operations guide should be specific to each individual system, and thus could comprise an entire book. In this chapter, we will only concentrate on a general procedure for one of the most problematic cases—namely, a cold well restart after an extended, unplanned shutdown.

A detailed operating procedure typically contains six components:

1. **Scope of operation**: Provides a high-level overview of the operation.
2. **Primary risks during operation**: These are brief descriptions of the risks that could occur during operation.
3. **Other safety, environmental concerns**: (not described here).
4. **Documentation, with references**: The operating procedures are written in modular format as guidelines that can be combined to make the system, with process and instrument diagrams (P&IDs), process flow diagrams (PFDs), specific to the system at hand.
5. **Prerequisites**: Conditions that need to be present in order for this procedure to be valid.
6. **Procedure**: A step-by-step operational procedure with monitoring remarks embedded, such as (1) how should the pressures, temperatures, and flows be changing, and (2) what to do if the changes are not proper.

In this chapter, we provide a general overview of the prerequisites and an example of a general operating procedure for startup of a cold subsea well and system.

7.5.1 Who Is the Customer?

The construction of an operating procedure requires substantial experience, not only of the equipment and process, but also with the individual offshore platform workers who will implement operating procedures. Such knowledge of those who will implement the procedure is vital to the success of any engineering process. Recent books have been published about offshore workers (Aven and Vinnem, 2007; Carter, 2007). However, with such a complex subject, the authors can only hope to paint the characteristics of platform operators with a broad brush.

Platform operators are typically highly intelligent and conscientious workers, motivated to ensure production in a safe and straightforward manner. Because they live with and operate offshore processes, operators understand such processes much better than an engineer, and should be consulted when writing operating procedures to ensure that they are practical.

Three characteristics of process operators particularly affect the way that an operating procedure is implemented:

1. Operators require procedures as simple and reliable as possible; more complex procedures have greater chances for errors.
2. Operators will work pragmatically to implement a written operating procedure, using the company's official procedure as authorization for their actions.
3. Operators prefer a single number (or two) to operate a system, rather than a graph or table. For example, the hydrate formation pressure at the ocean temperature is a better operator guide than a plot or table showing hydrate pressure as a function of temperature.

It is important to realize that operators have many other things in their charge, in addition to dealing with hydrates. The objective of maintaining safe production from the reservoir to the beach has a number of components that are too numerous to deal with here. The flow assurance component of the operator's job deals with more than hydrates. In the subsequent example, guidelines are given to deal with hydrates, the principal flow assurance problem in cold-well startups. However, potential flow assurance problems and guidelines should also be considered for (1) corrosion control, (2) paraffin/wax, (3) asphaltenes, (4) scale, (5) erosion, (6) sand, and (7) liquid slugging.

7.5.2 Writing an Operating Procedure

Before an operating procedure is written, three major steps must be performed:

1. Make a first approximation of the type of system, and whether hydrate protection is required.
2. Develop a strategy to mitigate the risks of the particular system.
3. Perform a detailed engineering assessment of key design parameters.

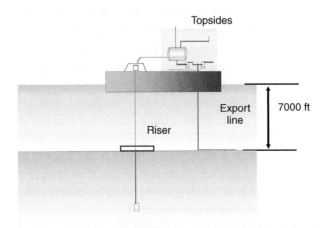

Perforations

Figure 7.3 Schematic of dry tree system in which the fluids are still warm at the platform choke, so that separation, drying, and dehydration can be done with a minimum of hydrate inhibition. Such a system is expensive, requiring short connections to the reservoir and fewer wells per platform.

The previous three steps are detailed in the following three paragraphs, before an example of an operating procedure is given:

1. First approximation of system type. In this initial portion of the design, the flow assurance engineer must assess whether a dry tree (Figure 7.3) or a wet subsea tree (Figure 7.4) will be used for each well, and whether injection of inhibition chemicals will be necessary for hydrate inhibition. This usually entails determination of the hydrate formation temperature at the reservoir and flowing pressures along the flowline, for startup, shutdown, and normal operations. Normally, flowlines are designed for oil systems such that flowline insulation prevents loss of heat, so that fluids arrive topside at temperatures greater than the hydrate formation temperature, under flowing conditions, without the injection of inhibitor. Hot dead oil circulation capability and inhibitor injection are always provided in the case of transient operations such as startup and shutdown.

2. Develop and document a strategy to meet the risks of a particular system. In this portion of the design, the flow assurance engineer determines the hydrate risks in all portions of the system, from the reservoir to the beach, and forms hydrate control strategies. Will sufficient water be produced to warrant a hydrate control strategy? Will the heat loss along the flowline lower the temperature below the hydrate

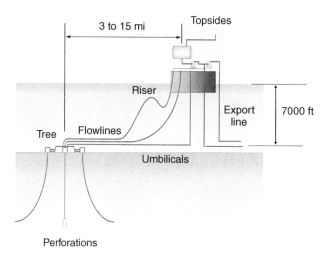

Figure 7.4 Schematic of wet tree design system in which more and longer flowlines are attached to a single platform. Wet systems generally require more hydrate inhibition precautions than the dry tree system shown in Figure 7.3.

formation temperature at flowing condition, or at stagnant conditions if the topsides choke is shut? Is the salt concentration in the combined produced and condensed water sufficient to naturally inhibit hydrates? Should the hydrate system be inhibited with methanol, monoethylene glycol, or low-dosage hydrate inhibitors (LDHIs) discussed in Chapter 5?

3. Perform a detailed assessment of key design and operating parameters. During this preparatory stage the engineer determines the lowest temperature that the systems will experience (usually the deep seafloor temperature of 39°F), using the maximum allowable operating pressure (MAOP) and a single worst-case fluid composition to generate a hydrate pressure–temperature curve. Heat transfer calculations are performed along the flowline, in flowing conditions to ensure that the fluids arrive topside above the hydrate formation temperature. After the flowing temperatures are determined, stagnant heat transfer calculations are made to determine the "no-touch" time, defined as the time after an unplanned upset, in which operations does not have to concern themselves with hydrate control. The cooldown time is also calculated for the wells, topside, and subsea components, defined as the time required from current operating temperatures and pressures before the line cools into the hydrate formation conditions. In addition, the

warmup time is provided as the time from the startup or restart conditions to the normal flow conditions at each flow rate. Typically, the operators monitor the temperatures and pressures at the wellhead, the subsea system, and at topside boarding, in addition to normal measurements on the platform regarding the separator conditions and composition of arriving fluids.

7.6 SAMPLE OPERATING PROCEDURE: COLD WELL STARTUP INTO COLD SYSTEM

Consider the diagram of a generic well system given in Figure 7.5. In this oil-dominated well example, six assumptions were made:

1. The well has been shut-in for a substantial amount of time (e.g., >4 weeks), so the cool-down has been sufficient to cause the well to come to the geothermal gradient temperature.
2. The hydrate formation temperature is known for the well production fluid at both maximum and flowing pressures.
3. The sub-surface safety valve (SCSSV in Figure 7.5) is placed at sufficient depth so that the SCSSV temperature is above the hydrate formation temperature, at the shut-in pressure.

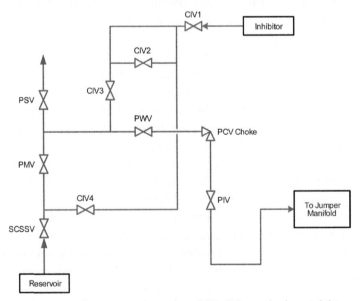

Figure 7.5 Generic well system with methanol (MeOH) as a hydrate inhibitor.

4. A hot, dead oil circulation temperature of 170°F is sufficient to maintain the system above the hydrate temperature when first production is added to the system.

5. When flowing at steady state, the reservoir temperature is sufficient so that insulation alone can prevent hydrate formation in all parts of the flowing system, without inhibitor.

6. A common thermodynamic inhibitor such as methanol (MeOH) or monoethylene glycol (MEG) is used until the system is ramped to flowing capacity sufficient to maintain all temperatures above the hydrate temperature, at which time inhibitor injection is suspended.

7.6.1 Pre-Startup Checklist

1. The tree and chemical injection valves (SCSSV, PMV, RSV, PWV, PCV, PIV, CIVs1,2,3 and 4 in Figure 7.5) should be closed.

2. The hot oil injection system should be checked to ensure that it can provide oil at 170°F through the flowline.

3. The function of the inhibitor injection pumps should be checked.

4. The storage of a sufficient quantity of inhibitor should be available.

5. The platform (topside) reception manifold should be aligned to the production test header.

6. The topside should be ready to receive product through the selected riser.

7.6.2 Restart Guidelines

The following guidelines should be adjusted to obtain a detailed procedure for each well.

1. Using the pigging pumps, circulate hot oil through the system until the system temperature is greater than 170°F, which will provide hydrate protection when the first production is started.

2. Equalize the pressures across the tree valves using methanol, and open valves SCSSV, PMV, and PWV.

3. Begin injection of methanol at the SCSSV and upstream of the choke (PCV) via opening valves CIV1, CIV3, and CIV4.

4. Begin injection of other inhibitors for asphaltene, scale, corrosion, and so on.

5. Open the PCV choke and ramp up the well to full production according to the well ramp-up schedule, obtained separately from engineering.

6. Stop methanol injection when the temperature upstream of the choke is >80°F, a temperature greater than the hydrate formation temperature for the well pressure.
7. Monitor the liquid rate and topside arrival temperature to ensure that it is above the hydrate formation temperature.
8. If the topside arrival temperature is too high (e.g., approaching 222°F) apply back pressure up to 600 psia using the topside choke. The wellhead choke PCV may have to be adjusted as well.
9. If slugging occurs topside, apply back pressure at the riser using the topside choke.

After the well has reached the target rate, and the temperatures have stabilized at less than 222°F at the end of the riser, but greater than 130°F at the subsea gauges, and inhibitor injection has ceased, then normal subsea operation has commenced.

7.7 RELATIONSHIP OF CHAPTER TO OTHERS IN THIS BOOK

Chapter 1 presents an overview of hydrates structures, followed by Chapter 2, which showed how hydrate plugs form. Chapter 3 provided safety guidelines for plug dissociation, and Chapter 4 provided plug dissociation details. Chapter 5 was written to describe the function of inhibitors, including naturally inhibited hydrate systems, in which no artificial hydrate inhibition techniques are needed. Chapter 6 provides one standard in which laboratory techniques are used to predicate kinetic inhibitor performance in flowlines. In this chapter, we provide methods for operating procedures for hydrate flow assurance.

REFERENCES

Aven, T., Vinnem, J.E., 2007. *Risk Management: With Applications from the Offshore Petroleum Industry*. Springer Press.
Carter, P., 2007. *Don't Tell Mom I Work on the Rigs: She Thinks I'm a Piano Player in a Whorehouse*. Da Capo Press.

CHAPTER EIGHT

Conclusion

Contents

This book provides heuristics or Rules of Thumb for:

When we consider how far hydrate flow assurance has evolved since 1934, when Hammerschmidt discovered hydrates in flowlines, there have been three eras:

1934–1944: Industry wide realization that hydrates cause blockages in hydrocarbon flowlines.

1944–1995: Generation of data and prediction methods to avoid hydrates by thermodynamic inhibitors, removal of water, or by increasing the line temperature, through such mechanisms of insulation, pipe-in-pipe, or heating.

1995–present: Movement to risk management, going beyond time-independent thermodynamic avoidance to time-dependent phenomena such as kinetic inhibition, stabilized (cold) flow, anti-agglomerants, and so on.

Newer era applications are frequently combined with the older era methods, for tools that address the new needs of deeper water, colder temperature, and better economics. An example of such combinations is when a classical inhibitor such as monoethylene glycol is used to suppress the thermodynamic formation temperature until the subcooling is sufficient to use a kinetic hydrate inhibitor such as polyvinylcaprolactam (PVCap).

This book was written by both industrial workers and academics to enable the practicing flow assurance engineer to determine how to prevent hydrates blockages, how to remediate them once they have formed, and how to safely deal with hydrates. The book gives references for the experimental details on which the concepts are based; to explicitly include

Natural Gas Hydrates in Flow Assurance
ISBN 978-1-85617-945-4
DOI: 10.1016/B978-1-85617-945-4.00008-X

all the details behind the concepts would make the book unwieldy for industry, either to read or to apply.

This concluding chapter provides summary heuristics, or rules of thumb, which are first expressed and detailed in the previous seven chapters, as guides for the engineer encountering or anticipating a hydrate problem. The combined set of heuristics comprises the current hydrate paradigm, or state-of-the-art, for hydrate flow assurance.

The rules of thumb are intended to be guides to action, but like most guides to action, there are exceptions to every heuristic that can be addressed first by reference to the chapter from which the guidelines are extracted, and subsequently by experience or by a hydrate specialist.

8.1 CHAPTER 1. BASIC STRUCTURES AND FORMATION PROPERTIES

1. Hydrates are crystalline compounds that require four conditions for formation: (a) water, (b) small hydrocarbons (<9 Å diameter), (c) low temperature (e.g., <80°F), and (d) high pressure (>200 psig). In the absence of any one of these conditions, hydrates will not normally occur.

2. Computer programs such as Multiflash®, PVTSim®, DBRHydrate®, HWHyd, or CSMGem are available to predict hydrate formation pressures and temperatures, as a function of hydrocarbon phase composition and thermodynamic inhibitor concentrations.

3. Because completely filled hydrates are 85 mol% water and 15% hydrocarbon, hydrates first form a film at the hydrocarbon–water phase interface, where sufficient quantities of both are available. There is insufficient water dissolved in hydrocarbon, or hydrocarbon dissolved in water, to provide substantial amounts of hydrate formation in either the bulk hydrocarbon or water phases.

4. When the hydrate film forms at the hydrocarbon–water interface, the solid hydrate barrier inhibits further hydrocarbon–water contact, slowing hydrate growth in the direction perpendicular to the film.

5. In an oil-dominated system, a hydrated film encapsulating water droplets in oil provides the first step to a hydrate plug, in which the amount of hydrate can be very small, as little as 4 volume % of the water present.

8.2 CHAPTER 2. HOW HYDRATE PLUGS FORM AND ARE PREVENTED

6. In offshore systems, hydrates form typically between the subsurface safety valve in the well and the beachhead. The task of a flow assurance engineer is to design systems to prevent hydrates, and to remediate them safely once they are formed.

7. Gas-dominated systems lose heat rapidly and the production fluid would cool to within the hydrate formation boundary, were not inhibitors (e.g., methanol and monoethylene glycol) injected downhole and/or at the wellhead to prevent hydrate formation. Thermodynamic inhibitor injection amounts can be determined using flow rates of the various phases, and the prediction programs in Item 2 above.

8. In oil-dominated systems, the higher heat capacity of the fluid, aided by insulation, preserves the high reservoir temperature above the hydrate formation point until the production fluid reaches the platform, where separation, and dehydration enables prevention before the fluids are compressed and shipped to the beach.

9. Hydrates can also form by sudden expansion of a gas, flowing through a restriction such as a valve in a fuel-gas takeoff line, so that inhibitor injection upstream of the valve is also common.

10. Due to the difficulty in describing mixed-phase hydrate formation, initial progress has concentrated on describing simpler end-members of dominant phases: oil-dominated systems, condensate systems, gas-dominated systems, and free water systems.

11. For oil-dominated systems in which all the water is emulsified, hydrates first form a film around each water droplet, and the hydrate-encrusted water droplets aggregate via capillary forces to form a plug. Over time the hydrate plugs solidify and become more difficult to remove.

12. In condensate systems, water droplet emulsification is not an option, so hydrates form on the walls, to decrease the diameter, resulting in eventual wall sloughing, and then jamming of hydrate particles to form a plug. If there is free water in the system, the plug forms as the system cools to hydrate formation conditions and remediation should be immediate. For dissolved (not free) water above the hydrate formation concentration, hydrate deposits are more uniformly distributed along the pipe wall.

13. For gas-dominated systems in which there is little condensate, hydrates commonly form just downstream of water accumulations, with a subcooling approximating 6.5°F. The gas bubbling through a water accumulation at a low point in the system will create substantial surface area when the gas exits the water, resulting in a hydrate formation point. The initial hydrated mass appears to be a honeycomb of hydrate-encrusted gas bubbles that collapse and harden over time. Hydrate deposits also form on water-wet walls of gas systems, similar to condensate systems.

14. Research is currently underway for systems with free water. However, initial results indicate that oil droplets in water form hydrates that are weakly attracted to each other, forming weaker agglomerates (by 75%) than do water-droplet hydrate aggregates in the oil phase in oil-dominated systems of Item 11.

15. In oil-dominated systems, newer plug prevention technology centers on the prevention of particle aggregation. New methods include anti-agglomerants, and stabilized (cold) flow in which all of the water is converted to hydrates, eliminating the possibility of strong capillary forces to aggregate hydrated particles.

16. Newer prevention techniques in gas-dominated systems focus on kinetic hydrate inhibitors, polymers that inhibit crystal growth and nucleation by adsorption on the surface of the hydrate crystal, forming a barrier to further crystal growth.

8.3 CHAPTER 3. HYDRATE SAFETY DURING REMEDIATION

17. Two properties of hydrates cause them to be a safety hazard:
    a. They are denser than the fluids with which they are associated, giving them momentum and subsequent impact when they are dislodged from the pipe wall.
    b. They contain significant amounts of gas, so that if they are dissociated in a confined space, they will generate significant amounts of pressure.

18. On depressurization of a pipe, hydrates dissociate radially, detaching from the wall due to a high temperature gradient from the ambient surrounding water to the interior of the plug.

19. With radial plug dissociation, if there is a pressure gradient, the dissociating plug can become a projectile when the plug detaches at the wall. Hydrate plugs, which have been measured at speeds as high as 270 ft/sec,

can rupture at a line obstruction or change in direction. Alternatively, the momentum of the speeding plug may cause pressurization of the line downstream, resulting in line failure.

20. Plugs should always be depressurized by two-sided depressurization if possible, allowing for the possibility that there may be several plugs in the line with pressure between them.

21. If single-sided dissociation is performed, it must be done with a great deal of precaution and forethought.

8.4 CHAPTER 4. INDUSTRIAL METHODS FOR HYDRATE PLUG DISSOCIATION

22. Each plug behaves differently and thought must be given to each individual remediation case when plugs form.

23. Traditional means of blockage location via back-pressurization (downstream) and relative pressure changes (upstream and downstream) have been supplemented by modern tools such as gamma-ray densitometers and hoop strain gauges.

24. There are recommended separate strategies for process flowlines, wells, risers, and equipment, as detailed in Chapter 4. Usually the recommended strategies are some variation of the following strategy for hydrates in a process flowline, perhaps with changes in order or degree of application.

   a. Consult a specialist.
   b. If a thermal method is included in the design, use the thermal method.
   c. If the thermal method is unavailable, use the chemical method.
   d. If the plug is closer than a safe distance from the facilities and a mechanical method is available, then use the mechanical method.
   e. If it is possible to depressurize the system below the hydrate equilibrium pressure, then use the pressure method.
   f. Nonstandard engineering solution is required.

25. Details of each method (thermal, chemical, mechanical, depressurization, etc.) are provided in Chapter 4.

8.5 CHAPTER 5. INHIBITOR MECHANISMS AND NATURALLY INHIBITED OILS

26. Thermodynamic inhibitors effectively remove water from participation in the hydrate structure by hydrogen bonding the inhibitor

(methanol and monoethylene glycol) to the water molecule. Salt inhibitors dissociate into ions that form strong coulombic attractions with water, preventing the water inclusion into the hydrate cages.

27. Kinetic inhibitors adsorb to the surface of a hydrate crystal, anchored by pendant groups along a polymer chain. The hydrate crystal is forced to grow around/between the polymer strands.

28. The object of kinetic inhibitors is to prevent hydrate growth to a blockage, for a time exceeding the residence time of free water in the pipeline.

29. With lower subcooling, hydrate growth can be blocked for a significant time, on the order of days by kinetic inhibitors. However, as subcooling becomes greater, nominally greater than about 10°F, the hydrate crystal will grow rapidly.

30. Anti-agglomerants (AAs) also adsorb on the surface of a hydrated droplet, and all of the water present is rapidly converted to hydrate.

31. AAs act so that the head group of the chemical attaches to the hydrate, while the long hydrocarbon tail maintains the hydrate particle as a dispersion in the liquid phase. In this way, hydrate particle loading in a liquid can be as high as 60% in the liquid hydrocarbon.

32. Agglomeration to blockage in oil systems is a function of the wettability of the hydrate particle and wall surface. If the hydrate particles and wall are oil-wet, the likelihood of hydrate particle attraction to other particles, or attachment to the wall, is much less than if the particles and wall are water-wet. In this way, AAs can be considered as chemicals that change hydrate particles from water-wet to oil-wet particles for lower agglomeration.

33. Naturally inhibited oils are oil-wet particles, due to the attachment of substances on the surface that change the particles from being water-wet to oil-wet.

34. The above blockage prevention techniques are based on chemistry for thermodynamic, kinetic, and AA inhibitors. However, shear is another significant variable that can help to prevent hydrate plugs. Agglomerated hydrate particles appear to have a yield stress like a Bingham plastic, before they are annealed to more unyielding, solid-like plugs.

35. Chapter 5 provides quantitative rules of thumb for the pressure drop required to restart a flowline as a function of the effective viscosity of the fluid.

36. Cold or stabilized flow occurs when the particles have all the water converted to hydrate, so there is little water available for capillary attraction to agglomerate the particles. This represents the "no-chemical" solution.

8.6 CHAPTER 6. CERTIFYING HYDRATE KINETIC INHIBITORS FOR FIELD PERFORMANCE

37. KHI behavior in field pipelines can be reproducibly and accurately predicted using flowloops of 0.5-in and 4-in diameters, as well as high-pressure autoclaves. High-pressure rocking cells appear to give unacceptably high data scatter.

38. Accurate prediction of KHI performance requires appropriate shear conditions (turbulent, flowing systems) and precise temperature control. Shut-in, stratified, and wavy flow regimes in lab tests result in large data scatter and generally do not agree with turbulent flow results.

39. The preferred method of detecting KHI failure in testing apparatuses is an increase in temperature with concomitant precipitous gas consumption. Gas consumption is seen as a pressure drop (in constant volume tests) or a volume loss (in constant pressure tests), depending on the experimental setup.

40. If KHI testing with hydrogen sulfide (H_2S) cannot be performed due to laboratory or materials constraints, carbon dioxide (CO_2) has been shown to give similar kinetic hydrate inhibitor hold times for the same subcooling and may be a good substitute for H_2S.

8.7 CHAPTER 7. OFFSHORE PRODUCTION OPERATING PROCEDURES FOR HYDRATE CONTROL

41. Deepwater oil systems are typically designed as a piggable loop, with active or passive heat management for hydrate control. Oil systems use inhibitors for startup, and usually operate without continuous inhibitor injection after the system is sufficiently warm.

42. Deepwater gas systems, with lower heat capacity than oil systems, are usually bare pipe using inhibitors rather than heat to manage hydrates.

43. In the procedures design of a subsea system, four questions should be asked and answered: (a) When and where are hydrates likely to form

in the production system? (b) What can be controlled to prevent hydrates from forming? (c) What are the monitoring points in the system that will give indication of hydrates? (d) If a hydrate plug forms in the production system, how can it be remediated? Typical answers to each of these questions are provided in Chapter 4.

44. It is vital to consider the customer and his/her background, so that the customer is intimately involved in each document. To provide hydrate control from design to operations, three steps are provided: (a) detailed design (engineering), (b) operating guidelines (engineering and operations), and (c) operating procedures (operations).

45. A detailed operating procedure is given for a cold-well startup into a cold line.

Appendix: Six Industrial Hydrate Blockage Examples and Lessons Learned

On March 10, 2009, the Colorado School of Mines (CSM) organized a workshop under the aegis of the Research Partnership to Secure Energy for America (RPSEA). The goal was to gather recent hydrate blockage examples from industry, at the Chevron Bellaire facility in Houston. Seven industrial examples of hydrate plugs were presented, but only six were approved for publication in this book. The three objectives of the workshop follow:

1. To generate a field hydrate blockage database to determine where, when, and how hydrate plugs formed in the field, and the most frequent and costly hydrate plugs
2. To move toward hydrate plug prevention and remediation standards
3. To provide data to improve lab and flowloop experiments, as well as models to locate plugs

These six case studies and the accompanying lessons learned at the appendix conclusion (Section A7) are provided to complement the four case studies detailed in Section 4.8. These ten case studies should be considered as updates to the 25 case studies of hydrate plug formation and remediation given by Sloan (2000, Appendix C).

In the industrial invitation soliciting hydrate plug case studies, CSM suggested 10 pieces of data required to thoroughly document a hydrate plug incident to enable modeling of formation and dissociation:

1. Specific plug formation details
2. Line topology
3. Hydrocarbon and water phase compositions and phase fractions
4. Pipeline diameter
5. Flowrates of oil, water, and gas
6. Pressures and temperatures with time
7. Plug indications (timing, location, etc.)
8. Initial liquid holdup

Natural Gas Hydrates in Flow Assurance
ISBN 978-1-85617-945-4
DOI: 10.1016/B978-1-85617-945-4.00014-5

9. Heat transfer coefficients (insulated, buried line, etc.), and ambient temperature.

10. Pre-incident operations should be added to the list, that is, opening of a well, chemical dosage modification, short restart, and so on.

Normally when hydrate plugs form, as indicated in case studies A3 through A6, it is difficult to obtain data such as those in the previous 10 suggestions, due to the emergency nature of the situation. The motivation is to safely remove the plug and to resume production as soon as possible. However, the first two field case studies were implemented deliberately in a considered manner to provide the best data for a gas system.

A1 WERNER-BOLLEY HYDRATE BLOCKAGE FORMATION FIELD TESTS

DeepStar CTR5209-1 Greg Hatton and Veet Kruka, 2002

The tests at the Werner-Bolley field were conducted to study the safe single-sided depressurization of hydrate plugs. The tests consisted of four intentional hydrate blockage formations in a gas condensate flowline. In addition to the blockage remediation, data were recorded during the blockage formation.

All four tests formed hydrates just after a low spot with likely water accumulation before an elevation increase over a mountain, when the subcooling of the line exceeded 6.5°F into the hydrate region. All four tests had similar plug formation behavior, described in four steps by Hatton and Kruka: (1) early-stage behavior characterized by a gradual increase of pressure; (2) middle-stage behavior, buildup, and collapse of pressure difference due to adhesion to the wall and subsequent release; (3) final-stage behavior, formation of a hydrate blockage that shuts in well flow; and (4) post-flow behavior, with limited gas permeability of the hydrate plug.

The Werner-Bolley field flowline is a 4-in diameter, 17,381-ft length line, which produced primarily gas, some condensate, and a little condensed water at rates of 4 MMSCFD, 100 BCPD, and 10 BWPD, respectively. Table A1 gives the composition of Werner-Bolley fluids.

The flowline elevation profile is shown in Figure A.1. The instrumentation (Sites 1 through 5) monitored pressure changes along the flowline

Table A1 Composition (mole %) of Werner Bolley fluids

N_2	CO_2	C_1	C_2	C_3	$i-C_4$	$n-C_4$	$i-C_5$	$n-C_5$	C_6	C_7^+	Total
0.43	1.51	78.2	13.59	3.61	0.39	0.69	0.25	0.21	.24	0.88	100

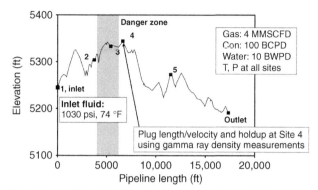

Figure A.1 Elevation profile of the Werner-Bolley flow line with locations of tempera-ture-pressure ports indicated. Shaded portion of plot shows most frequent site of plug formation.

with time and slugging of liquids and solids (via a gamma ray densitometer) at instrumentation Site 4. The temperature profile (Figure A.2) shows that the flowline enters the hydrate region just before Site 2.

Plug formation data were measured in January and February 1997. The four field tests with successful hydrate blockages were initiated by stopping the injection of methanol at the wellhead, pigging the pipeline with normal pro-duction, and then maintaining that production. This provided an initial con-dition without hydrate inhibition and with minor liquid holdup at test start.

Typical plug formation in the Werner-Bolley gas condensate tests occured in four stages. Figures A.3 and A.4 give the data from Tests 1 and 4. Tests 2 and 3 were similar.

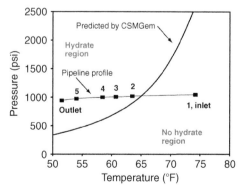

Figure A.2 The temperature-pressure profile of the flow line plotted on the hydrate stability curve (without methanol). Numbers denote positions of Bell Holes for P,T measurement as shown in Figure A.1.

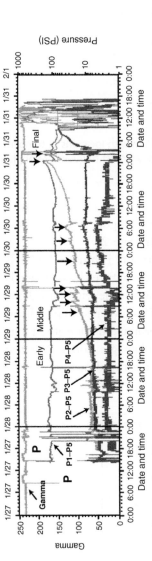

Figure A.3 Werner-Bolley data from Test 1, including pressure differentials (with respect to point 5) and gamma densitometer at point 4. P indicates pigging of the line to remove liquids (pig trip calculated at 45 minutes), including methanol (Deepstar CTR 5902-1, pg 3).

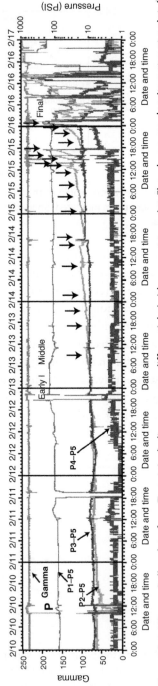

Figure A.4 Werner-Bolley data from Test 4, including pressure differentials (with respect to point 5) and gamma densitometer at point 4. P indicates pigging of the line to remove liquids (pig trip calculated at 45 minutes), including methanol (Deepstar CTR 5902-1, pg 3).

1. Early-stage behavior (ESB) was exhibited by a smooth, gradual pressure drop increase across the pipeline between points 2 and 5 (see denotations "Early" in Figures A.3 and A.4).
2. Middle-stage behavior (MSB) was exhibited by cycles of buildup and collapse of pressure drop as indicated by the downward pointing arrows in Figures A.3 and A.4). Partial blockages developed during the MSB may be transmitted downstream, shown by increases in downstream pressure differentials.
3. Final-stage behavior (FSB) showed formation of a hydrate blockage that shuts in well flow. This is shown by severely fluctuating pressure measurements.
4. Post-flow behavior (PFB) was exhibited by limited or no flow right after blockage. It is unclear whether the blockage strengthens with time.

The results of the tests are summarized in Table A2. The time to blockage was very similar for Tests 1 and 2, with a much shorter time for Test 3, and slightly longer for Test 4.

It was hypothesized that as gas emerged from the water accumulation, substantial surface area was created by the exiting bubbles. The resulting gas hydrates deposited on the walls, with growth eventually resulting in sloughing, and jamming as shown in Figures A.3 and A.4.

Table A2 Hydrate plugging time in Werner-Bolley field

Run	First observable pressure change (hr)	Time to first spike (hr)	Plugging time (hr)	Maximum pressure at the wellhead (psi)	Ground temperature (°F)
1	38	44	99	1030	44.6
2	36	48	98	1040	45.4
3	34	36	56	1120	45.4
4	72	110	143	>1200	43.8
Average	45	59	99	—	44.7

Lessons Learned from a Gas Field with Small Amounts of Condensate and Water

1. The plugs formed between Sites 2 and 3 in Figure A.1. This formation point was at the coincidence of a high water accumulation (before the flowline went over a mountain range) and a subcooling of 6.5°F (Matthews et al., 2000).

2. The tests additionally showed that it is possible to do hydrate dissociation via single-sided depressurization. See Hatton and Kruka (2002) for discussion of single-sided depressurization of plugs.

A2 THE TOMMELITEN GAMMA FIELD TESTS

Pål Hemmingsen

These tests, conducted in the spring of 1994, consisted of the rare instance when a company purposefully formed eight hydrate plugs in a service line to study the hydrate formation conditions. This test (Austvik, et al., 1995), combined with the Werner-Bolley test in Section A1, represent the only two deliberate formations of hydrates in flowlines in the literature.

A simplified schematic of the Statoil Tommeliten-Gamma tieback to the Edda platform is shown in Figure A.5a, in Spring 1994. The tieback consisted of two production lines, each 9 inches in diameter with a 6 inch diameter, uninsulated service line. The hydrate experiments were all carried out in the service line. According to Austvik et al. (1995), condensate and water mass fractions leaving the separator are 16 and 2 wt%, respectively; this field is considered as a gas field, with a little condensate, and less water.

While the elevation was very flat in that portion of the North Sea, the service line passed over several other pipelines between the subsea wellhead and the Edda platform, thus creating water accumulations, and additional dip sections in which hydrate chunks could jam. In the experiments the hydrate plugs consistently formed in the same subsea location, as indicated in Figure A.5b. The service line was produced into the test separator

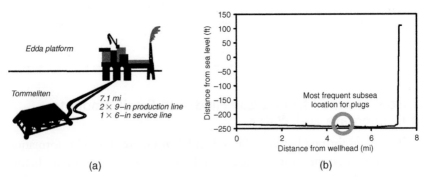

(a)   (b)

Figure A.5 (a) Tommeliten Subsea Tieback to the Edda Platform from Austvik et al. 1997 and (b) the Approximate Topology of the Tommeliten Service Line.

via a choke valve on the topsides. The following data were recorded during the field tests:

- The methanol injection rate
- The temperature and pressures in the well bore, service line riser and test separator
- Gamma ray densitometer readings of the riser topside
- The choke valve position
- The water and oil levels in the test separator
- The mass flow rates from test separator of oil, water, and gas

Figure A.6 shows the OLGA®-calculated water and oil accumulations where the elevation changes occurred in Figure A.5, with the hydrate formation occurring at the conjunction of (1) water accumulation and (2) adequate subcooling.

A total of eight experiments were performed. Hydrate plugs formed subsea in four of the eight experiments: namely Experiments 4, 5, 6, and 8 (Berge and Gjertsen, 1994). In the first experiment, the service line was restarted having been shut-in and filled with methanol for three months. A hydrate plug was present on restart, probably due to a leaky valve on the inlet allowing uninhibited water to enter the service line during shut-in.

In the second and third experiments, the service line was produced with no inhibitor injection. After a series of pressure and temperature fluctuations in

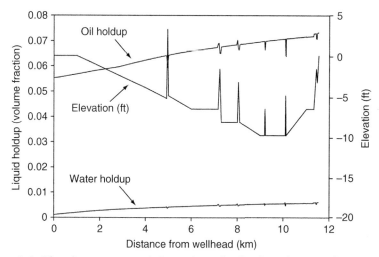

Figure A.6 Oil and water accumulations where the flowline elevation changes. Note that hydrates formed at about 7 and 8 km from the wellhead, with a conjunction of water accumulation and subcooling.

the riser, a hydrate plug formed topsides on the platform. In the fourth experiment, a depressurized line was restarted. A hydrate plug formed subsea. The position was estimated by raising the pressure of the service line from the platform side. In the fifth experiment, a hydrate plug was allowed to form during steady-state operation with induced failure of methanol injection. The plug formed in the same location subsea shown in Figure A.5b.

The sixth and eighth experiments involved the restart of the service line following a pressurized shut-in. In the sixth experiment, the line was uninhibited. In the eighth experiment, the line was underinhibited with methanol. Plugs formed and dissipated as indicated by pressure fluctuations and gamma ray densitometer readings. Eventually the line plugged subsea in 40 hr for the uninhibited experiment and 13 hr for the under-inhibited experiment. The seventh experiment was an attempted restart of the flowline after it was thought that the plug from the previous experiment had been removed by depressurization and methanol injection. The riser plugged almost immediately as the platform was restarted, attributed to residual hydrate in the service line from the previous experiment.

Lessons Learned from the Tommeliten-Gamma Field Experiments

The results from the eight field tests in the Tommeliten-Gamma field are remarkably similar to those from the Werner Bolley field, another gas field with limited condensate and little water, presented in Section A1.

In both fields, hydrates formed upon a conjunction of two conditions:
1. Shortly after water accumulated before a line elevation increase
2. When the subcooling was sufficient for forming hydrates

The time for hydrate plug formation and the length of the plugs were variable in each of these cases.

A3 MATTERHORN GAS EXPORT PIPELINE BLOCKAGES

Ronny Hanssen and Moussa Kane

The Total Matterhorn Tension Leg Platform in Mississippi Canyon Block 243 was installed in 2003 at a water depth of 2800 ft. It is designed with dry-tree production wells and there are two export lines: an 8-in oil line, and a ten inch gas line, producing through a 15-mi tieback to a junction with a 20-in trunkline, as shown in Figure A.7. Two incidents with hydrate blockage have occurred in the gas export pipeline (Kane et al., 2002).

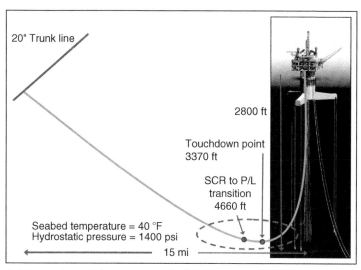

Figure A.7 Matterhorn 10″ gas export pipeline.

Hydrate Plug Following Hurricane Ivan

The gas export pipeline (GEP) was severely damaged as a result of Hurricane Ivan in September 2004. Mudslides caused the main 20-in gas trunkline to be displaced by 300 to 650 ft, which damaged the 10-in GEP at the tie-in point resulting in water ingress. During the repair pigging operation of the 10-in pipeline in March 2005, a hydrate plug formed. After consideration of the options to remove the plug, coiled tubing with methanol dissolution was selected. This action was successful to the SCR touchdown point, but not beyond. It was therefore followed by nitrogen displacement of the liquid head in the riser, while pressure was maintained to prevent early plug dissociation. Following liquid displacement, the line was depressurized for plug removal.

A timeline for restoration of production follows:

March 29–31: Coiled tubing mobilization; access through Y-plug in pipeline

April 1–8: Chemical dissolution (with methanol) of hydrate plug with limited success

April 9–14: Nitrogen offloading of the riser and depressurization to restore production.

Photographs of hydrate removed and at the pig receiver are shown in Figure A.8.

Figure A.8 Hydrate slurries, top two photographs. Bottom two photographs: (left) hydrate in pig receiver and (right) in barrel.

Hydrate Incident during Operation of the GEP

In January 2007, a hydrate blockage formed in the gas export line due to a glycol dehydrator re-boiler malfunction. The initial pressure increase was from 1320 psig to 1500 psig very rapidly (about 20 min) indicating a hydrate plug. The flow conditions of the line were such that the line was subcooled by 30°F at the most severe point, as shown in Figure A.9.

Due to the malfunction of the dehydration unit, high dewpoint gas was flowed through the GEP. Water accumulation, estimated to 0.3 bbl/d, took place at the lowest point of the GEP, creating conditions favorable to hydrate formation. The blockage, resulting in the observed pressure surge, occurred a day after the gas was back to a normal, low dewpoint. This delay points to a progressive aggregation of hydrate particles over a period of days, eventually leading to a relatively weak, unconsolidated plug. The plug was removed via depressurization and soaking with methanol. Coiled tubing was used to remove any liquid head above the plug and to place methanol at the plug face.

Matterhorn Lessons Learned

1. Coiled tubing can be very effective for both methanol placement at the plug face, and for gas lift to remove hydrostatic pressure on a hydrate plug, and thus enable depressurization resulting in plug removal.

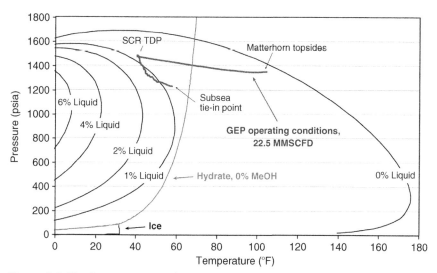

Figure A.9 The line pressure and temperature operating conditions imposed on a plot of the hydrate formation line (labelled Hydrate, 0% MeOH) and the vapor-liquid phase envelope (0, 1, 2, 4, 6% liquids). This plot shows the line to be sub-cooled almost 30°F at the SCR TDP, with approximately 1% hydrocarbon liquid condensed.

2. The dewpoint of the lowest temperature in the flowline must be considered for water condensation/accumulation and hydrate formation.
3. Due consideration must be given to the conditions of the entire flow path with respect to the phase envelope of the gas. At deepwater pressures and temperatures, condensate accumulation may occur and go unnoticed. This can present serious slugging risks, especially in the case of large fluctuations in production. In addition, liquid accumulation issues may appear or become more severe during the life of a field, as gas composition changes and/or production rate declines.

A4 ANADARKO INDEPENDENCE HUB HYDRATE PLUG REMEDIATION

Kevin Renfro and Nikhil Joshi

When hydrates form in a field, a pressure indicate may occur so that remedial action can be taken, as shown in the example of Figure A.10 for the Merganser Field.

The Anadarko Independence Hub normally operates the Jubilee #4 location at 4000 psia and a low temperature of 38°F. Unlike the Merganser field shown in Figure A.10, the Jubilee 4 system plugged rapidly, within 30 minutes of the first pressure increase, as shown in Figure A.11.

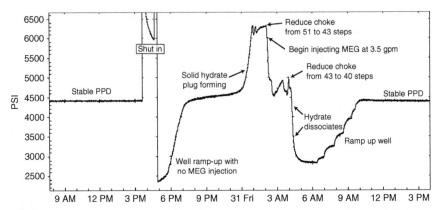

Figure A.10 In the Merganser Field above, the system was restarted about an hour after shut-in. The MEG inhibitor was inadvertently not begun with well restart. After restart, the pressure gradually increased on ramp-up, indicating the formation of a hydrate deposit. Upon the pressure buildup the choke rate was reduced and MEG injection was begun, resulting in a hydrate dissociation. After the pressure had reduced the system production was increased, this time with MEG injection and normal pressures, without hydrate buildup.

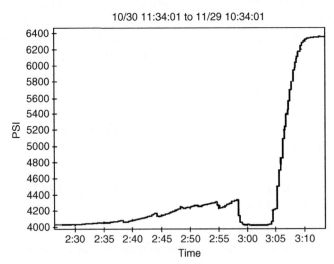

Figure A.11 Jubilee 4 downstream choke pressure (psia) versus time. Note that the pressure first increased at 2:35 pm and the line plugged at 3:05 pm.

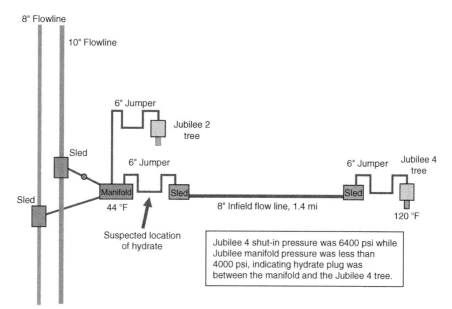

Figure A.12 Due to the pressure drop, it was determined that the most likely plug location was either the 6 inch jumper or sled or manifold, 1.4 miles downstream of the tree. The temperature dropped from 120°F to 44°F over this 1.4 mile length.

Due to the pressure drop, it was determined that the most likely plug location was either the 6 in jumper or sled or manifold, 1.4 mi downstream of the tree. The temperature drops from 120 to 44°F over this 1.4-mi length (see Figure A.12).

Hydrate remediation was achieved by pumping methanol over a 12-day period, both from the Jubilee 4 wellhead and from the Jubilee 2 wellhead. Approximately 6500 gal of methanol were pumped over this period. Figure A.13 shows the pressure over the period from November 6–15, 2008.

Table A3 provides a step-by-step account of the Independence Hub hydrate remediation.

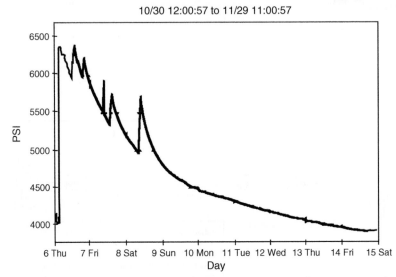

Figure A.13 Downstream pressure in Jubilee 4 versus time. Spike in pressure shows methanol injection points.

Table A3 Steps to Independence Hub Hydrate Remediation

1. Injected methanol periodically to Jubilee 4 tree and monitored pressure bleed off
2. ROV arrived to close sled valve on Jubilee 2 line
3. Began injecting methanol thru Jubilee 2 tree toward Jubilee 4
4. Total methanol injection estimated to be 5000 gal
5. Opened Jubilee 4 choke to flow a 10 MMCFD rate
6. Downstream (D/S) pressure at Jubilee 4 immediately pressured up to 6000+ psi
7. Pumped 1500 gal of methanol thru Jubilee 2 and Jubilee 4
8. Prior to restarting Jubilee 4, all wells were SI on that flowline
9. Opened Jubilee 4 choke 5 steps – estimated rate of 2 MMCFD
10. D/S pressure at Jubilee 4 decreased from 2603 psi to 2452 psi in 10 minutes
11. Increased choke to 36 steps (8 MMCFD rate)
12. D/S pressure at Jubilee 4 continued to fall to 2409 psi over 12 minutes
13. Increased choke to 38 steps (15 MMCFD rate)
14. D/S pressure at Jubilee 4 continued to fall to 2397 psi over 10 minutes
15. Increased choke to 40 steps (22 MMCFD rate)
16. D/S pressure at Jubilee 4 increased to 2403 psi over 10 minutes (friction)
17. Finally increased choke to 42 steps and maintained 31 MMCFD rate
18. Returned other wells to production

A5 ANADARKO'S BOOMVANG HYDRATE PLUG REMEDIATION

A diagram of the West Boomvang subsea layout is shown in Figure A.14. This field layout has the capability of cross-flow of fluids from lines D1 and D2. On February 24, 2004, well EB 686 #2 was brought online, flowing 23MMSCFD and 3100 BOPD with no water. On March 25 water breakthrough occurred, and on March 31, the production was 2MMSCFD, 47 BOPD with 4400 BWPD, overpowering the methanol injection system, resulting in a hydrate plug.

Figure A.15 shows the pressure and temperature transient on hydrate formation. Hydrate was dissociated by depressurizing the D2 flowline and confirmed by flowing gas across the pigging loop through the flowline back to the host. In the initial depressurization of the wellhead to 285 psig, 440 barrels of fluid (water, condensate, and methanol) were recovered. On the second and final depressurization attempt, 191 barrels of water were recovered, together with 81 barrels of condensate and methanol.

Flowline depressurization steps follow:

1. Shut-in EB 641 #1 and close both boarding valves. Displace methanol to both wells and flowlines through umbilical.

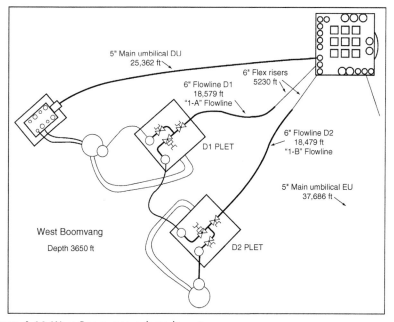

Figure A.14 West Boomvang subsea layout.

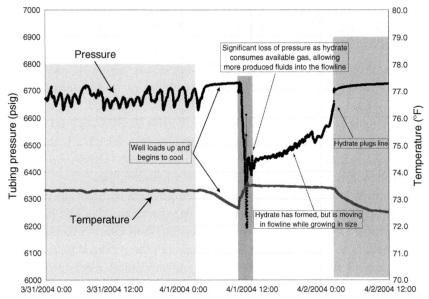

Figure A.15 Pressure and temperature transients on hydrate plug formation in D2 flowline.

2. Bleed residual pressure at surface and open D1 pigging valve.
3. Started methanol to EB 686 #2 and open D2 pigging valve.
4. Take flow from D2 flowline towards EB 686 #2 and into pigging loop and D1 flowline.
5. Shut pigging valves, open boarding valves and restart EB 641 #1 well.
6. Sweep D1 flowline of fluids accumulated during depressurization.

The next list provides the steps used to confirm hydrate dissociation.

1. Displace methanol into D2 flowline from riser and at subsea tree through umbilical.
2. Pump methanol into D1 flowline and close boarding valve. Allow EB 641 #1 to pressurize flowline to 75% of expected shut-in pressures at surface.
3. Restart methanol to EB 686 #2 and open both pigging valves.
4. Open D2 boarding valve, as pressure builds, and flow EB 641 #1 through D2 flowline.
5. Displace corrosion inhibitor and methanol into D2 flowline until recovered at surface.
6. Open D1 boarding valve and close pigging valves. Blow down D2 boarding valve and close.

A6 CHEVRON HYDRATE PLUG AT LEON CONDENSATE LINE

Douglas Estanga

The Chevron Leon platform delivers 100MMscf/d of gas and 20,000 bpd of condensate 30 km through an 8-in pipeline to an export facility platform, as shown in Figure A.16.

The sequence of events leading to the hydrate plug and its dissociation follows:

1. There was an extended shut-in for a time, greater than 1 month.
2. The multiphase flowline was restarted with injection of 50% methanol.
3. 10% of normal production was attained for 3 days.
4. The flowline was shut in again for a period of less than 1 month without methanol.
5. Upon restart, the flowline experienced a pressure buildup rate of 400 psi/min, indicating a plug.
6. There was a period of methanol injection and soaking before the plug dissociated.

One key to the hydrate plug formation was identification of the point of excess water holdup on shutdown, as shown in Figure A.17 for the two cases of liquid holdup with (left) normal and (right) 10% reduced production. This simulated profile in Figure A17 indicates that at reduced production, water accumulated in a low-lying portion early in the flowline, which may have diluted any methanol remaining in the line and thus promoted

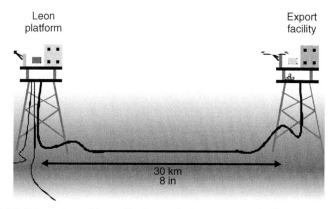

Figure A.16 Orientation of Leon Platform relative to export facility.

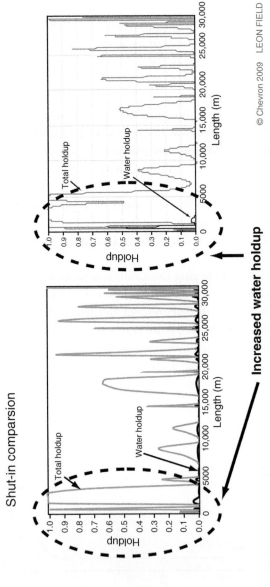

Figure A.17 Liquid and water holdup at shut-in for (left) full and (right) reduced production.

hydrate plug formation. Figure A.18 shows that the system temperature was below that of hydrate formation for the entire length of the line.

Figures A.19 and A.20 show simulations of the system using CSMHyK-OLGA® with relative viscosity, hydrate fraction, and pressure as a function of line length.

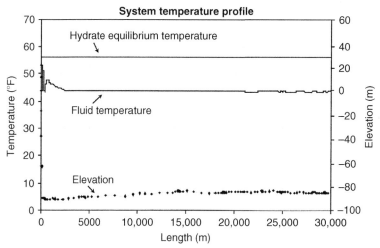

Figure A.18 Flowline topology (bottom), temperature (middle), and hydrate equilibrium temperature (top) as a function of line length. (*Courtesy of Leon Field,* © *Chevron, 2009.*)

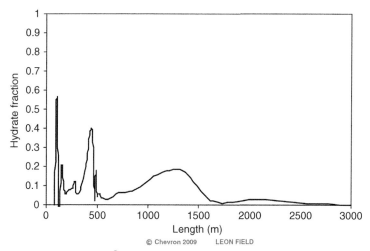

Figure A.19 CSMHyK-OLGA® prediction of increasing hydrate fraction at the point of maximum water holdup.

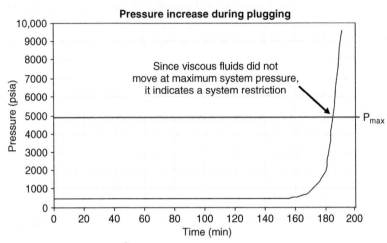

Figure A.20 CSMHyK-OLGA® simulation indicating hydrate plug (rapid increase in pressure) at point of maximum viscosity.

In the Leon plug and remediation, it was learned that the simulation of water holdup and low methanol concentrations were important ingredients to determine both the cause and location hydrate plug formation. The pug location was predicted accurately by integrating OLGA® and CSMHyK. Downtime was reduced in this case by methanol injection and soaking.

A7. LESSONS LEARNED AT MARCH 10, 2009, HYDRATE CASE STUDIES WORKSHOP

Norm McMullen

1. All cases cited during the March 10, 2009 meeting pertained to gas or gas/condensate. Oil case studies continue to elude us.
2. All oil systems that have plugged exhibit gas hydrate characteristics. It is suspected that all plugs result from free gas mixing with free water in the system, and oil has little to do with the plugging process.
3. For hydrate formation in gas lines with small amounts of condensate and free water, hydrates typically form just downstream of the free water accumulation, and the subcooling is 6.5°F or higher.
4. Methanol continues to be our "drug of choice," as it gives the best results. It was noted that in two cases the plugs were "permeable" to methanol.

5. System geometry dominates plug location. Most if not all plugs were located at or around a change in direction (jumpers, dips, elbows, etc.).

6. The causes of plugging were human error dominant. Secondary causes were beyond our control, such as hurricane damage or equipment failure.

7. It was notable that there were no comments or references to the time-dependent nature of plug characteristics, although we all suspect (based on laboratory observation) that plugs become "harder" and less permeable over time, due to a process of porosity destruction through compaction and geologic-like cementation.

8. In all cases it seems that locating plugs by conventional means is effective.

9. Flooded member detection (ROV-deployed Tracerco FMD Gamma Ray Densitometer) can be effectively used in certain circumstances.

10. Coiled tubing can be used effectively if it can reach the plug.

11. We now have the needed tools to effectively remediate everything that we have encountered to date.

REFERENCES

References for Hatton and Kruka
Hatton, G.J., Kruka, V.R., 2002. Hydrate Blockage Formation—Analysis of Werner-Bolley Field Test Data. DeepStar CTR5209-1 Final Report, August.

Matthews, P.N., Notz, P.K., Widener, M.W., Prukop, G., 2000. Flow loop experiments determine hydrate plugging tendencies in the field. In: Holder, G.D., Bishnoi, P.R. (Eds.), Gas Hydrates: Challenges for the Future, Ann. N. Y. Acad. Sci., vol. 912. 330–338.

References for Pål Hemmingsen
Austvik, T., Hustvedt, E., Meland, B., Berge, L.I., Lysne, D., 1995. Tommeliten Gamma Field Hydrate Experiments. Presented at the 7th International Conference on Multiphase Production. 7th–9th June 1995, Publication 14.

Austvik, T., Hustvedt, E., Gjertsen, L.H., Urdahl, O., 1997. Formation and Removal of Hydrate Plugs—Field Trial at Tommeliten, Gas Conditioning for Offshore. Presented at the 76th Annual GPA Convention, pp. 205–211.

Berge, L.I., Gjertsen, L.H., 1994. Field Test at Tommeliten Gamma Spring 1994, Statoil Internal Document. Doc. No. F&U-UoD/95015, Contract No. TGFLFHY2C, Filing No. 550.312/3.

References for Ronny Hanssen and Moussa Kane
Kane, M., Hanssen, R., Singh, A., 2008. Hydrates Blockage Experience in a Deep Water Subsea Dry Gas Pipeline: Lessons Learned. OTC 19634, Offshore Technology Conference, Houston.

INDEX

Note: Page numbers followed by *f* indicate figures and *t* indicate tables.

Printed and bound by CPI Group (UK) Ltd, Croydon, CR0 4YY

08/05/2025

01864850-0001